HOW RIVALRIES END

HOW RIVALRIES END

Karen Rasler
William R. Thompson
and
Sumit Ganguly

PENN

UNIVERSITY OF PENNSYLVANIA PRESS

PHILADELPHIA

Published by
University of Pennsylvania Press
Philadelphia, Pennsylvania 19104-4112
www.upenn.edu/pennpress

Printed in the United States of America on acid-free paper

2 4 6 8 10 9 7 5 3 1

A Cataloging-in-Publication record is available
from the Library of Congress
ISBN 978-0-8122-4498-4

CONTENTS

CHAPTER 1

The Problem of Rivalry De-escalation and Termination

The demise of the Cold War caught many, if not all, observers and participants alike by surprise. For much of the time between the end of World War II and the late 1980s/early 1990s, analysts and policymakers alike assumed that the East-West structural cleavage in world politics would remain unvarying. This cleavage was so paramount that it permeated and influenced world politics at all levels. In fact, for many observers every competition appeared, rightly or wrongly, as if it were a proxy struggle for the U.S.-Soviet rivalry. Then, abruptly, the central cleavage no longer existed. As a consequence, analysts and decision makers alike lost their conceptual anchor for deciphering how the world worked. The "world still worked," but a basic key to unlocking the secrets of how it worked had disappeared for good.[1]

In such periods of fundamental structural transformation, the usual response is to search for new ways to explain events and processes. Although the idea is hardly new, rivalry offers such an avenue. The history of international relations is replete with many examples of states engaged in long and intensive feuds. The Athenians and Spartans would have appreciated the idea of rivalry in the fifth century BCE. So would have the Romans and Carthaginians a few centuries later. The problem—perhaps because the phenomenon has been ubiquitous for so long—is that we have taken rivalries for granted. Therefore, the explicit study of rivalries as generic processes of conflict and cooperation is a relatively recent development. While there has been some effort at looking at nation-state dyads that have fought repeatedly prior to the end of the Cold War, there are still few clues as to why rivalries come and go. The unanticipated and abrupt end of the late twentieth century's "mother" of all rivalries—the Cold War between the United States and the Soviet Union—highlights this lacuna.

In this study, an expectancy theory pertaining to rivalry de-escalation and termination is advanced and tested.[2] The theory contends that rivalries end when four factors come together: shocks, expectational revision, reciprocity, and reinforcement. The examination of a single rivalry can help probe the utility of such a theory, but it can only be a starting point. More cases and more variation in outcomes are essential to testing the argument more comprehensively. Therefore, we examine ten rivalries: Egypt-Israel, Israel-Syria, Palestine-Israel, India-Pakistan, China–United States, China–Soviet Union, China-Taiwan, China-Vietnam, Thailand-Vietnam, and North Korea–South Korea.

This collection provides ample variance in rivalry behavior with four terminations, three significant de-escalations, albeit not permanently, and three continuing cases. Within these ten cases, there are also thirty-three episodes of attempted de-escalation with successful or failed outcomes in rivalry de-escalation/termination. For instance, the Egypt-Israel rivalry generates ten subcases of attempts at rivalry de-escalation; the Israel-Syria rivalry yields six; the India-Pakistan rivalry eight; and the North-South Korea rivalry three.

Moreover, some of these rivalries overlap partly as a consequence of their location in similar geopolitical neighborhoods. Hence, the interdependencies among these rivalries make it challenging to analyze them individually or as a combination or sets of cases. The concern is that a case-by-case analysis of these rivalries could overlook some theoretical breakthroughs that could be better observed when the interdependencies of these cases are taken into account via a more macro-investigation. Therefore, we propose to do both types of examination. In Chapters 3–7, we examine rivalries on a case-by-case basis. In Chapter 7, we investigate six rivalries as an ensemble within the Asian subsystem.

The Problem

Maoz and Mor (2002: 3) capture the heart of much of the motivation for rivalry analysis: "The belief in the pervasiveness of international conflict contrasts with the empirical record. A systematic survey of the last two centuries yields two seemingly striking facts. First, most states were relatively peaceful; only a small group of states was responsible for most interstate disputes and wars during this period. Second, these conflict-prone states did not fight randomly—they tended to fight each other. Consequently, a small group of dyads is responsible for a disproportionately large number of conflicts and wars."

More precisely, less than 1 percent of all the dyads in the past two centuries have been responsible for nearly 80 percent of the interstate warfare (Colaresi, Rasler, and Thompson 2007). While there is disagreement about how best to conceptualize these conflict prone states, scholars have begun drawing attention to the distinctive nature of rivalries and rivals—the dyads and states that tend to be most likely to become involved in international conflict.

What constitutes a strategic rivalry? Two important criteria are that states regard each other as competitive and as threatening enemies. States view other states as competitive when they are roughly in the same capability league. For instance, few scholars would argue that the United States and Mexico are able to mobilize capabilities at the same level. This disparity is due partly to the U.S. lead in resource endowment and partly to the seizure by the United States of a large proportion of Mexico's territory—which also ensured and enhanced the future lead of the United States over Mexico. Nevertheless, the United States and Mexico were once competitors of sorts, but they were not after 1848.

However, the stipulation that rivalry is restricted solely to those states that have symmetrical capabilities does not always hold. For example, the United States and Cuba have never been equal in strength, but, for a while, U.S. decision makers perceived Cuban foreign policy as highly threatening and Cuba as an enemy. Cuba reciprocated the favor. In recent years, the rivalry has atrophied, but neither side is willing to admit it.

States are viewed as enemies by the decision makers of a country if they have done physical harm in the past, or project some probability of doing such harm in the present or future. If they have already attacked or done damage to a state's interests, then the probability of inflicting more damage in the future is likely to be all the greater. If this perception is combined with competitive status, then there are four possibilities as delineated in Table 1.1. The most common situation is found in the lower right-hand cell: states that are neither especially competitive with one another nor explicitly threatening to each other. In the upper right-hand cell are states that are competitive but nonthreatening. Normally, this category would encompass states that might become political-military rivals sometime in the future or may have been rivals in the past but no longer perceive mutual threats emanating from each other. Examples of the former type are commercial rivals like Japan and the United States in the early 1990s. The latter type is exemplified by Britain and the United States or Britain and France after 1904, or France and Germany today.

Asymmetrical relationships, the lower left-hand cell, have already been discussed in the context of states that were not in the same league as their rivals but that acted temporarily as if they were. The more common situation is the genuine asymmetrical relationship in which a large and powerful state threatens a weaker neighbor and there is not a great deal that the weaker neighbor can do about it. Most recently, the Russian-Georgian relationship provides a good example. The point is that this type of structured dyad does not usually lead to a rivalry between the two states. At best, the weaker party can seek protection from a rival of the stronger party.

Asymmetrical dyads can become rivalries if the weaker side acts as if it is more powerful than its capabilities would otherwise suggest and the more powerful side reacts as if the weaker side is a threatening competitor. Examples would include Argentina-Paraguay in the mid-nineteenth century, Cuba–United States in the contemporary era, or, for a while, Cambodia-Vietnam. Ordinarily, one would not anticipate asymmetrical rivalries enduring for long periods of time without some sort of fundamental resolution of the dispute.

Strategic rivalries, combining similar levels of competitiveness and perceived threat, are located in the upper left-hand cell. They have not been common affairs. Of the many thousands of possible highly conflictual dyads, less than a couple of hundred have become strategic rivals. Yet their relationships tend to endure and escalate into hostility with some frequency.

Why rivalries begin, why they fluctuate in terms of their hostilities, and why they end are all interesting questions. In some respects, the question about the origins of rivalries may be the easiest to answer. Many rivalries—

Table 1.1. Competitors and Threats

	Threat	*Nonthreat*
Competitor	Strategic Rivalries	Competitor
Noncompetitor	Asymmetrical Relationship	Most dyads

approximately half of those that began after 1816—came into existence with the independence of one or more of the two feuding states. It is at this time that boundary disputes emerge and adjacent states that contain nationals thought to belong to or have some affinity with another state are subject to unwelcome internal intrusions. Another possibility is that two states in a region may be in contention over central leadership. All of these disagreements lead to threatening behavior and perceived threats that bring about rivalry relationships.

Determining why rivalries oscillate in hostility is undoubtedly the most challenging of the three questions posed above. Existing research has yet to develop a full understanding of rivalry behavior. Questions about how rivalries terminate, in comparison, are far more manageable. The issue is whether the effort to understand how and why rivalries end is worth attempting. We believe the answer is yes. Rivalries encompass dangerous processes. Nearly 80 percent of the wars since 1816 have involved confrontations between rivals. If rivalry is central to war making, then understanding rivalry termination processes should be critical to improving our knowledge about how to lessen substantially the probability of war. At the very least, it should help us to understand why wars break out when they do.

Currently, strategic rivalries are not proliferating at an exponential rate (Thompson and Dreyer 2011). On the contrary, they have become less common. Yet those that remain—rivalries involving the two Koreas, China and Taiwan, India and Pakistan, to name a few of the most salient ones—draw attention to the world's hot spots of interstate tension. As rivals acquire nuclear weapons or drag patrons and allies into confrontations, everyone has a stake in assessing what it might take to de-escalate these problem areas.

Moreover, there is always some potential for new rivalries to emerge or old ones to reemerge. World politics exhibits some continuing propensity for intensifying tensions between large, powerful states. An argument, for instance, can be made for the reemergence of both the Russian-U.S. and Sino-U.S. rivalries—albeit in more constrained hues than had been exhibited in their earlier manifestations. For a decade and a half, Russia had been too weak to seriously attempt regaining some of the Soviet Union's former status. That situation has not persisted. China and the United States have been rivals in the past and agreed that they both had incentives in the 1970s to de-escalate their conflict. But conditions have since changed. At the moment, China and the United States have declined to acknowledge formally the full extent of their incompatibilities over goals and preferences, but their interactions frequently flirt with strong overtones of rivalry intensification. Meanwhile, China also finds itself in protracted and at least intermittently hostile relationships with India and Japan. Therefore, the revival of terminated rivalries between China and Vietnam or Russia are not out of the question.

Rivalries are not the exclusive property of large, powerful states. Some regions remain highly rivalry prone. The Middle East has long been a leader in this respect. Even if Egyptian-Israeli hostility has been defused for some time, the rivalry between them persists. The defeat and occupation of Iraq in 2003 put some rivalries in abeyance (for example, Iraq-Syria, Iraq-Kuwait, Iraq–Saudi Arabia, and Iraq-Iran), but that is not likely to be a permanent situation either. Other rivalries in the region (such as Iran-Israel or Egypt-Sudan) possess some likelihood of becoming more intense than they have been. Parts of sub-Saharan Africa and Central Asia also appear to be potential places for rivalries to dominate the international political landscape.

While regions exhibit differences in the presence and intensity of rivalry relationships, we have no reason to expect that strategic rivalries will disappear in the near future. In fact, we anticipate that interstate rivalries will be renewed nearly everywhere as environmental problems (global warming and water), energy issues (petroleum), and/or economic difficulties (world depression) grow worse in the twenty-first century.

Therefore, since rivalries have been important in the past, continue to be important in the present, and are likely to remain important in the future, we need to decipher why they come and go. In this book, we concentrate on why rivalries end. When one side defeats the other side overwhelmingly, this question is not difficult to answer. In these cases, rivalries disappear because one or both sides are no longer competitors. But why do some strategic rivalries end even when neither participant in the rivalry has been defeated?

The Rivalry Termination Record

If strategic rivalries eventually terminate, then it follows logically that there are several possible ways in which this might come about. The two fundamental paths, however, can be derived from the definition of strategic rivalry. They include situations when 1) one or both states in the rivalry lose their competitive status, and/or 2) one or both states cease to be perceived as a threatening enemy. In the first causal pathway, the loss of competitive status occurs when 1) one side is defeated decisively and acknowledges defeat; 2) one side acknowledges defeat without war; 3) one or both sides experience political-economic exhaustion and/or intensive civil war.

The loss of competitive status is reasonably common in rivalry cases. Table 1.2 lists the rivalries that have terminated through decisive defeats or exhaustion in the past two centuries.[3] Roughly a third (48 of 139 or 34.5 percent) of the rivalries in the past two centuries have terminated when one side was forced to yield or withdraw from its dyadic contest. Table 1.3 lists another fifteen cases of rivalry that ended as a result of non-externally related war changes with one side acknowledging its inferiority to the other. Adding these cases (fifteen) to those in the last table (forty-eight) indicates that sixty-three, or 45 percent, of the termination cases are associated with

Table 1.2. Rivalries Terminated by Coerced Decision or Exhaustion, 1816–2010

Rivalry	*Duration*	*Rivalry*	*Duration*
Afghanistan-Iran II	1996–2001	Czechoslovakia-Poland	1920–1938
Angola-DRC	1975–1997	Egypt-Ethiopia	1868–1882
Argentina-Paraguay	1862–1870	Egypt-Ottoman Empire	1828–1841
Austria-France	1816–1918	Ethiopia-Italy	1882–1943
Austria-Italy	1847–1918	France-Italy	1881–1940
Austria-Ottoman/Turkey	1816–1908	France-Vietnam	1858–1884
Austria-Russia	1816–1918	Germany-Poland	1933–1939
Austria-Serbia	1903–1920	Germany-Russia II	1890–1945
Bolivia-Paraguay	1887–1938	Germany–United States I	1899–1918
Brazil-Paraguay	1860–1870	Germany–United States II	1933–1945
Britain-Burma	1816–1826	Guatemala-Nicaragua	1855–1907
Britain-Germany I	1896–1918	Hungary-Rumania	1918–1947
Britain-Germany II	1934–1945	Indonesia-Netherlands	1951–1962
Britain-Italy	1934–1943	Italy-Turkey	1884–1943
Britain-Japan	1932–1945	Japan-Russia	1874–1945
Bulgaria-Yugoslavia	1878–1955	Japan–United States	1898–1945
Burma-Thailand	1816–1826	Lithuania-Poland	1918–1939
Cambodia–S. Vietnam	1956–1975	Mexico–United States	1821–1848
Cambodia-Vietnam	1976–1983	Mozambique-Rhodesia	1975–1979
Chad-Libya	1966–1994	Poland–Soviet Union	1918–1939
China-France	1844–1900	Saudi Arabia–Yemen I	1932–1934
China-Japan I	1873–1945	Thailand-Vietnam I	1816–1884
Czechoslovakia-Germany	1933–1938	Vietnam–S. Vietnam	1954–1975
Czechoslovakia-Hungary	1920–1938	Yemen–S. Yemen	1967–1990

Table 1.3. Rivalries Terminated by One Side Accepting Inferiority without Being Forced to Yield by Defeat in Warfare, 1816–2010

Rivalry	*Duration*	*Rivalry*	*Duration*
Argentina-Brazil	1817–1985	Guatemala-Mexico	1840–1882
Austria-Prussia	1816–1870	Iran–Ottoman Empire	1816–1932
Britain-China	1839–1900	Iran-Russia	1816–1828
Britain-Russia	1816–1956	Mozambique–South Africa	1976–1991
Chile–United States	1884–1891	Ottoman Empire–Russia	1816–1920
China-Vietnam	1973–1991	Spain–United States	1816–1819
Colombia-Nicaragua	1979–1992	Yemen–Saudi Arabia II	1990–2000
France–United States II	1830–1871		

endings along realist lines. In short, one side came to the realization that it lacked the ability to continue the competition. Whether it dropped out willingly or otherwise is not our immediate concern.

The main alternative pathway—the downscaling of enemy threat perception—can be realized on one or both sides through such noncoercive processes as

1. shifts in strategic priorities (as in emphasizing domestic development versus external competition);
2. changes in domestic leadership (presumably more critical in situations in which the leader is a principal source of foreign policy orientations);
3. alterations in regime (and thereby leading to a redefinition of who constitutes friends and enemies);
4. negotiations in a mutual effort to lesson tension, hostility, and threat without any structural changes in the adversary's orientation.[4]

Table 1.4 lists the seventy-six cases that fall within the noncoercive pathway to rivalry termination. These cases are grouped together without making a distinction about the reasons for how and why enemy status was revised.

Meanwhile, Table 1.5 summarizes all the 139 cases along the two causal pathways of rivalry termination: competitive status changes and erosion of enemy status. Forty-five percent of these cases were related to the loss of competitive status, while 54 percent of the cases are associated with the loss

Table 1.4. Rivalries Terminated Noncoercively, 1816–2010

Rivalry	*Duration*	*Rivalry*	*Duration*
Afghanistan-Iran I	1816–1937	Ghana–Ivory Coast	1960–1970
Albania-Greece	1913–1996	Ghana-Nigeria	1960–1966
Angola–South Africa	1975–1988	Ghana-Togo	1960–1995
Argentina-Chile	1816–1904	Greece-Ottoman/Turkey I	1827–1930
Bahrain-Qatar	1986–2001	Greece-Serbia	1879–1954
Belize-Guatemala	1981–1993	Guatemala-Honduras	1840–1907
Bolivia-Peru	1825–1932	Guinea Bissau–Senegal	1989–1993
Britain-France II	1816–1904	Honduras-Nicaragua I	1895–1962
Britain–United States	1816–1904	Honduras-Nicaragua II	1980–1987
Bulgaria-Greece	1878–1947	Hungary-Yugoslavia	1918–1955
Bulgaria-Ottoman/Turkey	1878–1950	Indonesia-Malaysia	1962–1966
Bulgaria-Romania	1878–1945	Iran-Iraq I	1932–1939
Burkino Faso–Mali	1960–1986	Iraq–Saudi Arabia	1946–1958
Burundi-Rwanda	1962–1966	Israel-Jordan	1848–1994
Chad-Sudan I	1964–1969	Italy-Yugoslavia	1918–1954
Chile-Peru	1832–1929	Jordan–Saudi Arabia	1946–1958
China-Russia I	1816–1949	Kenya-Somalia	1963–1981
China–Soviet Union II	1958–1989	Kenya-Sudan	1989–1994
China–United States I	1949–1972	Kenya-Uganda	1986–1995
Colombia-Ecuador	1831–1919	Libya-Sudan	1974–1985
Colombia-Nicaragua	1971–1990	Malawi-Tanzania	1964–1994
Colombia-Peru	1827–1935	Malawi-Zambia	1964–1986
Costa Rica–Nicaragua I	1840–1858	Mauritania-Morocco	1960–1969
Costa Rica–Nicaragua II	1948–1992	Mauritania-Senegal	1989–1995
Costa Rica–Panama	1921–1944	Morocco-Spain	1956–1991
Dominican Republic–Haiti	1845–1893	Ottoman Empire–Serbia	1878–1918
DRC-Rwanda	1996–2009	Oman–S. Yemen	1972–1982
DRC-Uganda	1996–2009	Rhodesia-Zambia	1965–1979
Ecuador-Peru	1830–1998	Rwanda-Uganda	1999–2009
Egypt-Iran I	1955–1971	Saudi Arabia–Syria	1961–1970
Egypt-Jordan	1946–1970	South Africa–Zambia	1965–1991
Egypt-Libya	1973–1992	South Africa–Zimbabwe	1980–1992
Egypt-Saudi Arabia	1957–1970	Sudan-Uganda I	1963–1972
Egypt-Syria	1961–1990	Tanzania-Uganda	1971–1979
El Salvador–Guatemala	1840–1930	Thailand-Vietnam II	1954–1988
El Salvador–Honduras	1840–1992	United States–Soviet Union	1945–1989
France-Germany II	1816–1955	Soviet Union–Yugoslavia	1948–1955
France-Russia II	1816–1894		
W. Germany–E. Germany	1949–1973		

Table 1.5. Summarizing General Types of Rivalry Termination, 1816–2010

Competitive status	*Enemy status*
Coerced surrender (*N*=48)	Noncoercive (*N*=76)
Acknowledgment of inferiority (*N*=15)	

or erosion of enemy status. In fact, seventy-six of the cases represent rivalry termination that occurs through noncoercion. Of the sixty-three coercion-related cases of rivalry termination, one or both rivals lost their ability to continue the competition and were forced to yield, either by their adversary or by circumstances. These latter cases are less interesting to study because the reasons for rivalry termination are obvious and less puzzling.

However, the remaining seventy-six cases depict rivalry terminations that occurred without either side being able to claim an obvious victory. The reasons for their endings are less obvious and therefore more interesting to explain. There is no certain guarantee that the advent of internal shifts in domestic conditions or the presence of negotiations will automatically lead to rivalry termination by both sides. Once a rivalry forms, both sides develop highly negative images of their adversaries. Decision makers expect their rivals to engage in undesirable activities. They have little reason to trust one another. A domestic change in priorities, leadership, or regime may signal genuine changes in foreign policy intentions by one party. But the other side may choose to ignore signals or view them as part of a deceptive strategy. As for negotiations, they can drag on forever without any breakthrough unless both sides are genuinely interested in de-escalation. The obvious question is whether we can develop a relatively parsimonious explanation for occasions in which domestic priorities/leader/regime changes and/or negotiations are more likely to lead to rivalry termination.

The approach adopted here involves the construction and testing of a theory that focuses on decision makers changing their strategies toward rivals.[5] The probability of strategic changes are predicated on antecedent changes in expectations about rivals, external and internal shocks, the presence or absence of policy entrepreneurs (leading decision-makers with specific agendas to change the way governments operate), third-party pressures, reciprocity, and reinforcement. Not all of these factors are necessary to bring about rivalry de-escalation and termination.[6] Nor are they all likely to be of equal significance. Existing theories about rivalry termination are limited to a large extent by their uncertainty about how these factors interact consistently across cases as well as their inability to map all the possible combinations of factors that were present as rivalries ended. Therefore, one of our goals in this study is to explore how these factors do interact and which combinations seem to be the most important ones for explaining how rivalries end.

Once we elaborate our theory of rivalry termination, we will test its applicability in a number of cases encompassing Eurasia. Specifically, we will examine three Middle Eastern rivalries, one South Asian rivalry, and six East and Southeast Asian rivalries in Chapters 3–7. The ten rivalries under investigation generated thirty-three cases of attempts—some successful— at significant rivalry de-escalation—all of which are also examined in Chapters 4–7. Chapter 8 assesses the aggregate outcomes of the rivalry termination cases in terms of our expectancy theory, which is elaborated in Chapter 2. We find ample support for our explanation of rivalry de-escalation and termination. The pattern is quite clear: shocks, expectancy revision, reciprocity, and reinforcement are necessary ingredients in bringing about reduced conflict in antagonistic dyads. Policy entrepreneurs and third-party intervention may facilitate movements toward less rivalry conflict, but they are neither necessary nor sufficient.

CHAPTER 2

The Evolution of Expectations and Strategies

How do rivalries de-escalate and eventually terminate? Some international relations scholars argue that these outcomes depend on favorable background conditions that encourage adversaries to believe in the settlement of their outstanding disagreements.[1] Shifts in these background conditions create moments of opportunities or "ripeness" when peacemaking efforts have a chance to have positive outcomes. In other words, critical turning points occur when decision makers have the opportunity to pursue alternative strategies of conflict resolution. Which factor or combination of factors introduce critical choice points and why is left unclear.

We argue that an expectancy framework is best positioned to address these questions, because rivalries are assumed to be dynamic processes that are subject to modification in the context of environmental change. Moreover, this approach treats rivalries as protracted conflicts that are sustained by the strategies and policies of adversaries and their institutions. Over time, these strategies are reproduced in a routine fashion that produces inertia in the rivalry relationship. To modify the relationship, this inertia must somehow be overcome or upset.

Rivalries de-escalate or end when adversaries assume new interpretations, understandings, and expectations of their opponents. Such innovations can occur when environmental crises or shocks bring about the realization of other conceivable expectations. Shocks, for instance, can threaten the political survival of actors who, in an effort to secure their positions, question the viability of existing conflict patterns and repertoires of state action. Or, shocks can facilitate the rise to power of new decision makers with different expectations. As shocks hasten the reevaluation process, actors move from one set of collective understandings to another. If the direction of the shift produces negotiations, then information about these collective understandings will be exchanged through diplomacy. That is, leaders will communicate their new understandings to their opponents, and such communication will, hopefully, reinforce the process of negotiation and conciliation.

In the pages below, a model of expectancy revision is summarized and elaborated. Special attention is placed on the potential impact of shocks on altering decision makers' expectations and, ultimately, moving their strategic choices in the direction of de-escalation and rivalry termination.

A Model of Shocks, Expectancy Revision, Reciprocity/ Reinforcement, and De-escalation

In our model expectancy revision plays a key intervening role between shocks and rivalry de-escalation and/or termination as do policy entrepreneurs, reciprocity, reinforcement, and, to a lesser extent, third-party pressure. At the core of rivalries is intransigence that results from the entrenched expectations that each side has of the other in terms of perceived intentions, tactics, and past actions. Shocks can lead to the transformation of intractable conflicts if they encourage adversaries to reevaluate their prior expectations. But there is no anticipation that shocks suffice to bring about change.

Equally important, shocks can have escalatory or de-escalatory effects. Their impacts are highly contingent on timing, context, and changes in leaders' expectations. For instance, Jervis (1997: 126) suggests that environmental stimuli that set off positive feedback at one point in time can also set off negative feedback at another as the state of a system changes. This possibility makes it difficult to posit a deterministic outcome. Consequently, shocks are at best necessary but insufficient triggers for de-escalation processes. Despite this variability, it is reasonable to hypothesize that shocks and expectancy changes are more likely to bring about de-escalation when they converge with four other variables: policy entrepreneurs who have sufficient political control to overcome internal commitments to older strategies, external third-party pressure, reciprocity, and reinforcement. The latter variables are noteworthy because they do more than just aid the initiation of de-escalation. They sustain the process through continued positive feedback.

The connections between shocks, policy entrepreneurs, reciprocity, and third-party pressure are based on the following propositions. First, actors develop strategies to deal with (internal or external) rivals on the basis of expectations concerning what other actors are likely to do and what their own capabilities to deal with rivals are likely to be. Expectations are predicated in part on what rivals have done in the past and are currently doing. Strategies, meanwhile, are plans for coping with adversaries' behavior. They are developed in response to the interpretations of an opponent's behavior, within the context of other environmental considerations (for example, other threats and opportunities, domestic coalitions, allies, capability calculations, and competing demands for resource allocation). Expectations, strategies, and behavior, thus, are not identical phenomena. One does not necessarily translate automatically into another, but the three should be related, with expectations producing changes in strategies that subsequently lead to changes in policy and behavior. During protracted conflict, as expectations among the adversaries become entrenched over time, strategies and policy actions become routinized. The result is inertia and deeper intractability. When this routine is disrupted, however, expectations, as well as strategies and actions, may be characterized by uncertainty, thereby, encouraging some actors to reorganize their previous ways of perceiving and interacting.

Second, shocks are transitional situations that can instigate a major period of change in adversarial relations by altering key expectancies. Since shocks are not always certain to alter expectancies, the extent to which they do so ultimately depends on how actors interpret them. If shocks corroborate current orientations, then actors are unlikely to adjust their strategies and actions accordingly. However, should shocks disconfirm their expectations, actors are more likely to reassess the validity of their orientations and, if necessary, abandon or revise them. This reassessment is by no means guaranteed if, for instance, actors in an effort to protect their original expectations explain away inconsistent information. In short, shocks can be transforming events only if they cause adversaries to realize that their past strategies cannot triumph or they cannot gain more by continuing them and an accommodative strategy promises to offer a better alternative (Kriesberg 1998: 217).

Shocks can be either exogenous or endogenous, although neither is assumed to have a more important impact than the other. Exogenous shocks emerge from an environment that is external to the protracted conflict (for instance, changes in the international or regional distribution of power, new threats, war), while endogenous shocks (such as economic crises) occur within the domestic contexts of the adversaries. Shocks that have the greatest impact will be those that alter how threatening the adversary appears and/or how efficacious one's own side is likely to be in coping with the adversary. The more entrenched the expectations and the deeper the strategic inertia, the stronger the shocks must be in order to tip expectations in a new direction.

Next, policy entrepreneurs on one or both sides can be crucial in bringing about de-escalation. As strategic pioneers, entrepreneurs are willing to take on and sometimes overturn policy monopolies—the dominant political understandings or orthodoxies about a policy issue—in this case, the conditions surrounding negotiations and cooperation with an adversary. Because policy monopolies are reinforced by existing institutional structures (which limit access to the policy process) and powerful supporting ideologies that reside within these institutions (Baumgartner and Jones 1993), entrepreneurs must first consolidate their political influence by effectively eliminating or removing their internal opposition and promoting the role of like-minded reformers.

Shocks can figure prominently in this entrepreneurial process. Critical events (or potentially significant transforming events) can be policy windows that allow proponents of change to assert their political leadership by advancing new alternatives to old problems. Policy entrepreneurs who favor moderation enhance their chances of implementing strategic shifts if they are able to associate hard-line policies with past foreign policy failures and convince their constituencies that learning from these failures requires a more moderate approach (Evangelista 1991).

Another consideration is the role of third-party pressure on the principal adversaries. External mediation can be helpful in bringing about de-escalation, but it is unlikely to be successful in the absence of either expectancy revision by one or both adversaries or policy entrepreneurs. Shocks may play an important role here, too. External mediators are well aware of the fact that certain periods are more propitious (or ripe) for bringing about de-escalation than others. Therefore, shocks can stimulate or renew third-party attempts in coordinating peace efforts, encouraging innovative initiatives by adversaries, providing incentives for settlements, and contributing to the implementation and durability of the agreements (Kriesberg 1998; Hartzell 1999; Walter 1999). Although third-party pressure appears to be neither a necessary factor nor a sufficient one for de-escalation, its presence in the context of an appropriate shock is more likely to reduce intractability than at other times.

Finally, reciprocity and reinforcement are essential in bringing about de-escalation (Goldstein and Freeman 1990; Lebow 1995; Goldstein et. al. 1998; Kriesberg 1998). Concessions made by one side without appropriate responses from the other side cannot be expected to contribute to a de-escalatory process. Cooperation from one's adversary reinforces the expectations about future cooperation and strengthens the shift toward moderation. Moreover, continued reciprocity between the adversaries is critical to supporting the transition process beyond the initial political breakthrough. The question is when reciprocity will bring about de-escalation. Since shocks converge with the rise of strategic innovations, the advent of policy entrepreneurs, and the renewal of external mediation, it follows that reciprocal behavior will affect de-escalation similarly.

The Theory

Our theory can be summarized in the following way. Decision makers create assumptions about their own preferred foreign policy behavior (*strategies*) on the basis of perceptions (*expectations*) of external threats, the capabilities of their enemies, and the resources available to cope with external threats. Over time, the expectations formed about external rivals become entrenched. Changing these entrenched expectations may require some combination of radical changes in the environment (*shocks*), new decision makers with control over their governments and less allegiance to old expectations (policy *entrepreneurs who occasionally develop consolidated political positions*), and encouragement from external patrons (*third-party pressures*). Once new strategies begin to be experimented with, intransigence upon the part of the enemy (a lack of *reciprocity*) and/or the failure of the new strategies to achieve a de-escalation in hostility (a lack of *reinforcement*) are likely to lead to an abandonment of the strategic experimentation and a relapse to earlier strategies.[2]

At the same time, it seems probable that none of these factors are sufficiently powerful on their own to bring about the termination of a rivalry. Subject to further analysis, we think shocks, expectational change, reciprocity, and reinforcement are absolutely essential to de-escalation. Shocks interrupt inertia or at least may do so. Without expectational change about threats and resources (their adversary's and their own), it is hard to imagine decision makers proceeding toward some type of accommodation. In the absence of any reciprocity for concessions made by one side, it is also difficult to imagine de-escalatory processes continuing for very long. For the new, less adversarial relationship to persist, reinforcement is necessary to maintain normalized interactions.

Entrepreneurs with consolidated control of their regimes are likely to be quite significant in coordinating major changes in strategy. It is even possible that they are necessary as well, but it is also possible that changes in rivalries can be attained without extraordinary leadership efforts if the shocks and environmental changes are sufficiently strong in their own right. Similarly, third-party involvement could also be facilitative, but, unlike the case for policy entrepreneurs, we are fairly sure that third parties are not necessary. In the right circumstances, rivals are fully capable of winding down their own hostile relationships.

Assumptions

We make a number of assumptions that are best stated explicitly rather than left to the reader's imagination. One assumption is that actors rarely function as unitary decision makers in pursuit of fixed national interests. Various domestic groups, including important governmental agencies, have different goals and attempt to influence governmental agendas and behavior. When feasible, groups will seek to capture their government in order to monopolize these aspects. Government leaders who wish to stay in power must attempt to juggle these internal demands within the context of external demands on governmental behavior. One of the principal ways in which this can be done is to organize and maintain a coalition of domestic groups. Maintaining a coalition requires the pursuit of interests and agendas that appeal to the coalition in question. Hence, foreign policy will reflect to variable degrees the identity of the ruling domestic coalition, assuming of course that there is one. As the identity of the coalition undergoes change, so too will the interests and agendas that are pursued.

Actors are certainly not hyperrational. They operate with imperfect information, have hazy ideas about their own values and preference schedules, and do not necessarily weigh all options and then proceed with the least costly, most advantageous alternative. Instead, foreign policy formulation and execution is likely to be a process of trial and error in which policies emerge after a number of experimental probes in different directions. Just which direction will be privileged is not always clear. However, actors in leadership positions do monitor their environment for threats and opportunities. They also attempt, variably, to respond to perceived changes in threats and opportunities.

Actors categorize other actors as either competitors or noncompetitors. Competitors are further distinguished according to whether they represent some possibility of physical attack. While noncompetitors may also be threatening, external competitors that are potential attackers are considered strategic interstate rivals. Actors develop strategies to deal with rivals (either internal or external) on the basis of expectations concerning what other actors are likely to do and what their own capabilities to deal with rivals are likely to be. As we have noted, expectations are predicated in part on what rivals have done in the past and are currently doing, and in part on what nonrivals have done and are doing. These expectations are not easy to change once they have developed because people are reluctant to revise cognitive filters for interpreting stimuli.

Given these assumptions, we can express our theory in the following propositional format:

1. Actor expectations, strategies, and behavior tend to be characterized by inertia and are subject to repeated shocks of varying magnitude. In the abstract, any shock may be viewed as an opportunity for expectational and strategic revision. However, inertial constraints usually are difficult to overcome. *To tip expectations (and thus strategy and behavior) from one established routine to another requires fundamental alterations in expectations.*
2. *In turn, fundamental alterations are made more probable by major shocks that force actors to reevaluate the accuracy and appropriateness of their existing expectations and associated strategies.*
3. The types of shocks that have the most impact are those that alter how threatening the adversary appears and/or how efficacious one's own

side is likely to be in coping with the adversary. *Shocks that decrease perceived threat or the actor's own capabilities are likely to lead to reevaluation and de-escalation/termination. Shocks that increase perceived threat or the actor's own capabilities are likely to reinforce expectations and lead to intensified rivalry.*

4. *The more entrenched the expectations are or the greater the strategic inertia is, the greater or more multiple are the shocks needed to influence expectations.* No matter how great or multiple the shocks, though, there may still be variable lags between revisions in "sticky" expectations, strategies, and behavior.
5. Shocks must be interpreted. The same shock may be viewed positively or negatively vis-à-vis prevailing expectations. *As a consequence, a shock alone is not likely to be sufficient for expectational revisions unless the outcome of the shock completely eliminates the competitive ability of one or more adversaries.*
6. Possibly important are changes in leadership (for one actor or both) that go beyond mere personnel changes. *New leaders, especially ones committed to developing support for new policies, may also be committed to changing external relationships and to facilitating attempts at rapprochement. To be most effective, the leadership changes must also remove or neutralize sufficiently opposition by governmental and domestic elites to revisions in expectations, strategy, and behavior. That is, they must achieve consolidation to be effective.*
7. Third-party pressure for revision may reduce the resistance encountered by domestic policy entrepreneurs, but it is unlikely to be successful in the absence of fundamental revisions in expectations and leadership. *Third-party pressures thus may be facilitative but are unlikely to be either sufficient or necessary.*
8. *A further necessary ingredient in the revision process is the adversary's reciprocation at some level for any initial concessions made as part of an overture toward strategic and behavioral revisions.*
9. Once expectations (and strategies and behavior) have tipped from one regime or routing to another (or are in the process of tipping), it cannot be assumed that the new relationship is stable. On the contrary, new relationships emerge haltingly. *Consistent reinforcement of expectational revisions is necessary to prevent lapses back to the previous relationships still favored by historical conditioning.*

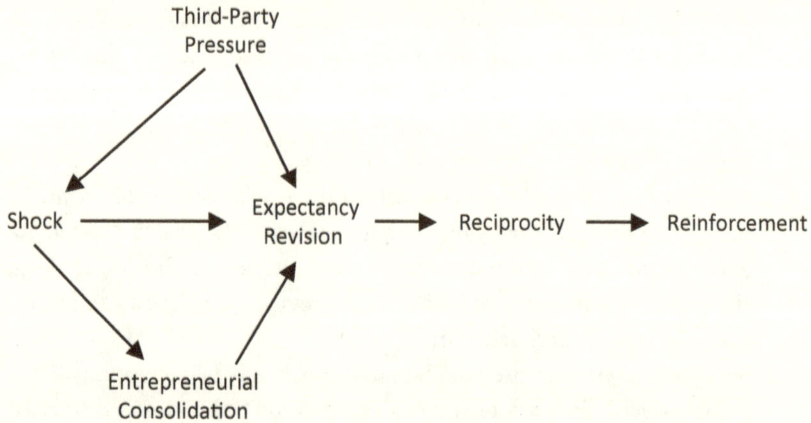

Figure 2.1. An expectancy model for rivalry de-escalation and termination. The number of arrows shown is considered a minimal number for maintaining simplicity. Arrows might easily be drawn linking entrepreneurial consolidation and reciprocity/reinforcement. Double-headed arrows, no doubt, should be drawn in the connections among expectancy revision, reciprocity, and reinforcement. Third-party pressures might also be manifested in terms of entrepreneurial consolidation, reciprocity, and reinforcement.

Before we proceed to test this theory, sketched in outline form in Figure 2.1, something more needs to be said about delineating the shocks and other variables that are so central to our argument. We also need to consider what case or cases will best serve our theory-testing goals.

Shocks and Other Variables

The cases of rivalry termination that have been described as the easiest to explain are those involving decisive defeats, exhaustion in civil war, or voluntary surrenders due to the recognition that a competition has become too asymmetrical to continue the contest. These situations, to say the least, represent extremely strong shocks for decision makers. If the nature of the shocks is overwhelming, then the shocks can suffice to terminate rivalries. Germany's defeat in World War II in some respects represents a mixture of decisive defeat, exhaustion, and newly created asymmetry vis-à-vis its former rivals. The defeat was so complete that it ended the possibility of rivalries with the Soviet Union, the United States, and Britain—the main agents of the defeat. Interestingly, though, it did not suffice to end the possibility of the Franco-German rivalry. Whether this was due to Germany's potential for revival, France's limited role in defeating Germany, the history of oscillations in the Franco-German relative positions, the long duration of their antagonism, their geographical proximity, or some combination of these factors is not of immediate concern. The point is that even very substantial shocks do not always eliminate expectations about the probability of renewed antagonism. Even tsunami-level shocks occasionally need assistance in bringing about change. Shocks, for that matter, may well work in the opposite direction. Depending on how they are perceived, shocks can reinforce mutual hostilities—thereby making the rivalry an even more entrenched behavioral pattern.

Clearly, shocks are critical to this interpretation. They may also be more complex than they seem at first. The main type of shock that tends to reinforce rivalry is a defeat in war that stops short of eliminating one of the two actors as a competitor. Put another way, a defeat in war may strongly suggest that one of the two actors is no longer competitive with its adversary, but this objective fact may not be internalized by decision makers on the defeated side. We have all sorts of decision-making pathologies in the foreign policy literature to explain why this large-scale misperception may occur.

The types of shock that are most pertinent to de-escalation/termination are four: shifts in external threat, changes in regime orientation/strategies, change in competitive ability, and domestic resource crises. As suggested by Figure 2.2, the four are hardly independent of one another. Rather, they tend to feed into one another. The two most straightforward types of shock are the changes to external threat and competitive ability. Decreased threat and deteriorated capability (relative or absolute) are singled out by the theory as the most potent types of changes because they go directly to the heart of rivalry expectations about threats from competitive enemies. Alterations in the perceived level of threat and/or the perceived ability to compete should (but not necessarily will) influence rivalry calculations. Nor should it be surprising to find that changes in competitiveness and threat are often reciprocal. Less competitiveness can lead to reduced threat perceptions. Reduced threat perceptions can lead to reductions in military preparations that could diminish the ability to compete militarily.

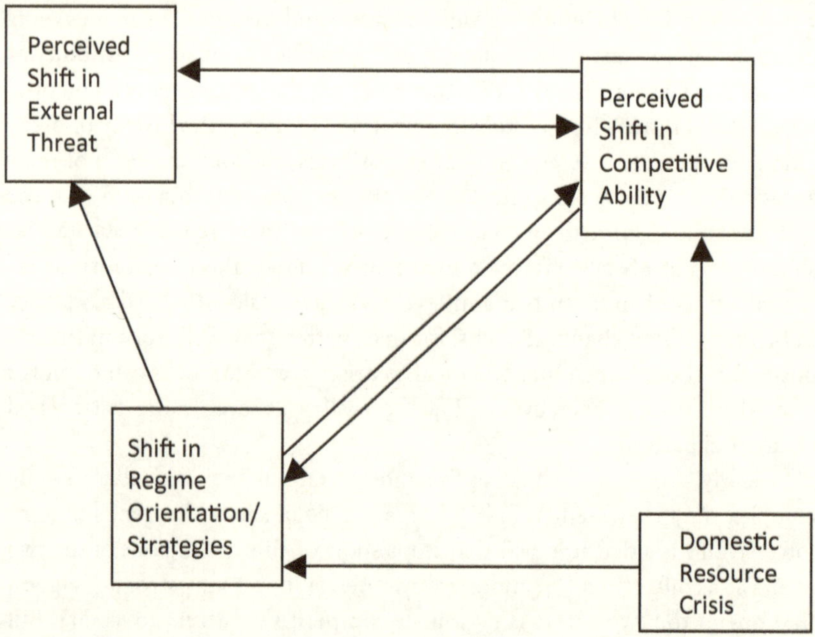

Figure 2.2. Shocks and their potential for interaction.

Changes in regime orientations or strategies can alter both the sending of threatening signals and the perception of threat by the other side. States in which foreign policies are dominated by an individual can assume a much different profile once that individual is removed from the scene. Democracies tend to respond favorably to autocracies that become more democratic. Or, in the absence of regime/personnel changes, a dramatic change in state priorities—as in reductions in militarization and foreign policy activity in favor of focusing more heavily on domestic economic development—can also affect levels of perceived threat. Similarly, domestic economic crises can both sap capability and lead to changes in strategy and regime orientation.

The strong probability of overlap in the implications of shocks is suggestive. Not only are some shock impacts likely to be more influential than other types—as the theory suggests—but the more compounding the effects are may also make some difference. The four types depicted in Figure 2.2 can work together as in a benign positive spiral of hostility de-escalation.

For instance, serious economic crises and competitiveness problems might lead to changes in regime strategies and threat perceptions. But the spiral could also work in the other direction as well. A vastly improved economic resource base and increased competitiveness can encourage more ambitious foreign policies and increases in threat perceptions on the part of other states. Shocks, therefore, can be both an intermittent change in the environment and also a motor for changing international relations.

We need to be attentive to the effects of shocks and whether they occur in compounded clusters. Yet shocks are only one ingredient in the dynamic chain of linked processes sketched in Figure 2.2. The main theoretical questions are whether expectations are as critical to rivalry de-escalation processes as the theory suggests, whether shocks are needed to tip expectations, and what mix of other factors are most important to attaining rivalry termination. Reciprocity and reinforcement would seem to be absolutely critical to defusing protracted conflicts. Can the same be said of third-party pressures? Are new leaders with radical, change-seeking agendas necessary?

But what should we be looking for in terms of the other variables? Expectancy revisions are about changes in how decision makers view the adversary. By definition, rivals are seen as competitors posing sufficient threat to be branded as enemies. The revisions that we should therefore be most interested in are situations in which rivals are viewed as projecting more or less threat or becoming more or less competitive. Less threat and/or less capability, the theory suggests, should be conducive to rivalry de-escalation and termination.

A shorthand term for the type of entrepreneurial leaders that we need to be looking for does not readily come to mind. We should be alert to new leaders—new at least to the position of head of government—who come into office with the notion that fundamental change in the relationship with the adversary in question would be advantageous. The theory does not argue that established incumbents cannot entertain or initiate rapprochements. The argument, rather, is that new leaders have an edge in introducing fundamental changes in foreign policy. Established incumbents tend to be too closely linked to older expectations and strategies. Yet the advantaged leader must not only be new and committed to changing rivalry relations, she or he must also possess sufficient control of the government to overcome inertia and built-in resistance to change from within the governmental bureaucracy and the larger society in which the government operates. Thus, we are looking for new, change entrepreneurs who are also sufficiently consolidated in political terms to be able to introduce policy innovations.

Third-party pressures refer to other states that have some interest in seeing (or thwarting) a de-escalation in a rivalry. Our approach is dichotomous. It is either present or absent. Clearly, third-party pressures can be intense, mild, or something in between. Delineating how much pressure is exerted and, equally important, how continuously it is exerted goes beyond our present concerns. We are only interested in whether third-party pressures seem to be necessary—or merely part of the story from time to time.

Reciprocity means that one side's concessions are met with some form of positive response. It may be symbolic, verbal, or material. The absence of any evidence of reciprocity should doom any attempt at rapprochement. Reinforcement means that once a rivalry relationship has changed its form, both sides need to continue behaving in a nonantagonistic mode. Just how long we should insist on reinforcement is arbitrary. Short-lived de-escalations are not impressive. Extremely short-lived cooperation may be difficult to identify as indicating any real change in the relationship. Even intense rivals can cooperate on some issues. The question is whether the relationship has genuinely changed. We need some minimal benchmark. Five years of consistently altered behavior is one minimal expectation.

It is certainly conceivable that every case does not require each and every one of the six factors that we are singling out for attention. Shock, expectancy revision, and reciprocity may suffice in one instance. Shock, new policy entrepreneurs, and third-party pressures may work elsewhere. Alternatively, new policy entrepreneurs may be highly significant in one case and only marginally present in another. We need to remain open that there may be multiple pathways—as in Lebow's argument outlined in the appendix—to de-escalation. While the specific combination of factors may be variable, however, we do contend that the presence of four of the factors (shock, expectational change, reciprocity, and reinforcement) greatly increases the probability of a significant de-escalation. We do not ignore the contribution of policy entrepreneurship or third-party pressures. We think, however, that they are less contributory to rivalry de-escalation in the aggregate than our four core variables.

In summation, then, Table 2.1 provides a quick list of things to look for in our cases.

Table 2.1. Operationalizing the Expectancy Theory

Variable	*Definition and indicator*
Shock	Abrupt and significant change in operating environment that is brought about by internal or external events and processes: 1. Shift in perception of threat—new enemies emerge or old enemies become more acute threats; 2. Change in regime orientation/strategies—major ideological/institutional transformation as in an autocracy becoming a democracy or a switch in strategies from unqualified intransigence to selective (or greater) accommodation; 3. Change in competitive ability—defeat in war, severe internal turmoil, obsolescence of military weaponry, failure to keep up in economic development; 4. Domestic resource crises—economic depression/stagnation, famine, environmental deterioration, serious energy shortages.
Expectancy revision	Perceptions that a rival is less (more) threatening or less (more) competitive and therefore less (more) threatening; or, that one's own side is seen as less competitive vis-à-vis the rival's capabilities.
Consolidated and new policy entrepreneurs	Principal decision makers, relatively new to office, who are unusually receptive to the idea of altering external relationships. The decision makers must also consolidate their control of the government's bureaucracy sufficiently so that they can expect to overcome internal resistance to foreign policy change.
Third-party pressures	Existence of encouragement toward (away from) rapprochement by states outside the rivalry—usually restricted to major power pressures.
Reciprocity	Exchange of symbolic and material cooperation and/or positively responding to opening overtures by one side of a rivalry to the other.
Reinforcement	Once a de-escalation/termination has occurred, the rivals (former or otherwise) continue to operate more or less at the new lessened level of hostility for a period of at least five years.

Cases

More information on how these factors have played out in specific cases is needed both to probe further the plausibility of the theory's line of argument—that is, whether we are isolating the right factors—and to see whether anything important may be missing or exaggerated. Yet there are major limitations in examining one case at a time. A general theory is unlikely to fit each and every case exactly. The question should be whether individual cases seem to approximate the general argument—with due allowance for variable weights of different factors in different cases and the possibility of different combinations of influences leading to similar outcomes.

But another problem associated with individual cases is that they have only one outcome. Either they de-escalated/terminated or they did not. Fairly static outcomes do not always help to reveal the dynamics that preceded them. Whatever the case, there is also a tendency to credit whatever seemed to precede the actual outcome with some causal efficacy. Outcome variance helps alleviate some of these problems. That is, if we have cases in which termination/de-escalation did and did not occur—or occurred to varying degrees, temporarily or seemingly permanently—we put ourselves in a better position to isolate correctly the relative role of alleged inputs to the processes at hand than if we have no variance. Of course, that is all variance does: it facilitates assessing the causal effects of multiple factors. It does not guarantee that the assessment will proceed accurately.

Variance can also become something of an analytical problem if the number of cases under investigation balloons beyond what is manageable. While manageability is always a subjective issue, the goal of probing a theory's utility suggests a more limited number of cases. More definitive tests with higher *N*s can come later. The question then is how best to set up an examination with limited *N* and variance? One way is to look for "natural experiments." What this means in terms of interstate rivalries is that they are often bundled in complex nests as opposed to free-standing and isolated, dyadic situations. The Middle East offers a good if very complex example. Israel has a rivalry with Egypt, but it also has or has had rivalries with Jordan, Syria, Iraq, and Iran. Egypt also has had or has rivalries with Jordan, Syria, Iraq, and Iran. Jordan and Syria have been rivals. Iraq and Iran and, probably, Syria and Iraq continue to be rivals. We could complicate this Mashriqi nest even further if we also introduced overlapping Gulf, Maghrebi, and Northeast African rivalry nests into the picture. The point here is that some regions offer outcome variance while, at the same time, help to restrict the sometimes excessive heterogeneity introduced by large *N* studies.

Taking on all the Middle East nested rivalries at one time would be quite a challenge in its own right. But even if we did focus exclusively on the rivalries of one region, we could never be sure that the analysis was not strongly biased by factors possibly unique to the region.[3] Our preferred approach, therefore, is to work on a larger pan-Eurasian, geographical palette. We examine ten rivalries. Three are from the Middle East (or Southwest Asia) and one is from South Asia. Two are from Southeast Asia, one is from Northeast Asia, and three might best be said to be located in East Asia.

The Middle Eastern rivalries (Egypt-Israel, Israel-Palestine, and Israel-Syria) are all interrelated in the sense that what happens in one has some effect on how the other two operate. As we shall see, though, there are definite limits to their interdependencies. The six rivalries in eastern Eurasia (the two Koreas, China-U.S., China–Soviet Union, China-Taiwan, China-Vietnam, and Vietnam-Thailand) are also intertwined, and, in some respects even more closely than the Middle Eastern cases. We think it makes a good deal of sense to follow a study of the Sino-U.S. and Sino-Soviet rivalries with an examination of adjacent rivalries that seem to have been affected by the termination of the local major power feuds. In this respect, the sequential termination of major regional rivalries constituted shocks to neighboring rivalries with varied effects. One reason for these reverberations is that Russian and Chinese relations with the two Koreas, Taiwan, and Vietnam were all affected by the end of the Cold War in general, and the associated winding down of the Sino-Soviet rift.

States highly dependent on external patrons are likely to be affected heavily if the patrons lose incentives to maintain their patronage. That does not mean, however, that dependent states will react identically to such an external shock. Nor do we wish to focus exclusively on client-patron type problems—hence our examination of the two major power rivalry terminations but also the Indo-Pakistani rivalry, which certainly possesses significance in its own right but also highlights the limitations of patronage dependency. Both Pakistan and to a lesser extent India have at times been dependent on some of the actors involved in the major power triangle consisting of the United States, the Soviet Union/Russia, and China. While these dependencies have influenced the Indo-Pakistani rivalry, they do not appear to be among the most important drivers of the continuing South Asian confrontation.

As this discussion hints at, it is awkward to examine rivalries in Eurasia without giving some attention to adjacent rivalries. If their interdependencies, hinted at in Figure 2.3, demand simultaneous attention, then it is also highly beneficial that rivalries in Eurasia offer attractive rivalry outcome variance.[4] One finds no change in the cases of Syria-Israel, India-Pakistan, and China-Taiwan, some varying amounts of de-escalation in the cases of Egypt-Israel, Palestine-Israel, and the Koreas, and outright terminations in

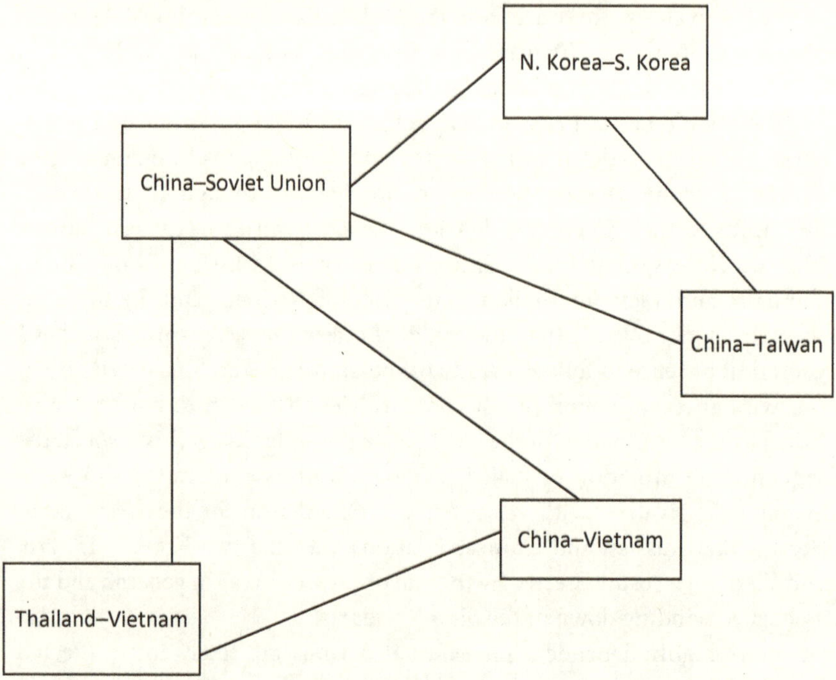

Figure 2.3. Interdependencies between rivalries.

the cases of China–United States, China–Soviet Union, China-Vietnam, and Thailand-Vietnam.[5] The Egyptian-Israeli, Israeli-Syrian, Indo-Pakistani, and Korean cases, moreover, offer something of a variance bonus with histories of multiple attempts to de-escalate. This variance bonus nicely expands the number of cases that we can examine. Depending on how one counts, this gives us a relatively manageable set of some thirty-three cases with which to examine the expectancy theory.

At the same time, we must acknowledge that not all the cases are fully equivalent. The complications of Egyptian-Israeli interactions make it more convenient to look at eras of different heads of government as the unit of analysis. In the Israeli-Syrian, Indo-Pakistani, and Korean cases, the cases are identified by episodes of attempted de-escalation by one or both parties within the rivalries' history. In the remaining instances, the rivalries themselves, or selected intervals, provide the boundaries delimiting the cases. We recognize the awkwardness of mixing ostensibly disparate units of analysis but think that our aggregation of these cases is defensible.

We do not give equal attention to all the rivalries examined in this study. Equal attention, from our perspective, would involve treating all rivalries exactly the same. If we examine one rivalry's full history, then we would need to do the same for the other nine. We do not proceed on this basis mainly for two reasons. One is that the ten rivalries are not equally interesting. A rivalry between China and the United States, for instance, is inherently more significant than one between Vietnam and Thailand. We can say this because a rivalry between China and the United States can have a great deal more impact than can one between two relatively small Southeast Asian states. Because of the greater potential impact of the major power rivalry, moreover, we also know more about the circumstances (or at least think we do) than we do about many minor power rivalries. Giving more weight to the big cases is thus natural.

But a second reason for giving the ten rivalries differential treatment is that we are interested primarily in end games. How do rivalries terminate? It would obviously be a great error to only examine rivalries that have ended. We very much need variance in outcomes to attempt some evaluation of what matters in rivalry de-escalation. Yet we do not necessarily need maximal variance—that is, we do not need to devote equal time and resources to the full story of every rivalry. From time to time, we can focus on episodes that might have led to de-escalation and termination but did not. We do this particularly in the Israeli-Palestinian and Korean cases, but also to a large extent in the Sino-U.S. and Sino-Soviet cases. As long as we include cases that led to termination or no change, there should be no suggestion that we are "rigging" the examination deck by exercising some selectivity in what we choose to analyze.

Nonetheless, we must remain sensitive—as should the reader—to the limitations of our approach. One good example pertains to shocks. We think shocks are highly important and we will attempt to relate them to rivalry outcomes—either as de-escalatory or escalatory—in our cases. Yet what we will continue to lack is a chronology of all the shocks experienced by decision makers. If we had such a series, then we could better estimate the association of shock with rivalry behavior. How many times did shocks occur without some change in rivalry behavior? How many times did some change in rivalry behavior occur without the occurrence of shock? Without such information, we are on shaky ground in asserting some relationship between shock and consequent behavior. Shaky ground though it may be, it will not keep us from making relationship assertions. To proceed otherwise would lead to analytical paralysis. But we and the reader need to keep in mind that more rigorous examinations in the future will be needed to nail down these asserted relationships.

Given our strong interest in the circumstances surrounding rivalry termination, it should come as no surprise that we feel no obligation to write extensive descriptions about the course of each rivalry. We are not historians. Nor are we writing rivalry histories. In several cases, very comprehensive descriptive accounts already exist. Not only do we feel no compulsion to compete with these already existing works, we have often relied upon them to assess the applicability of our own arguments.

Of course, circumstances might be even more ideal if there were less variance exhibited in the non-outcome dimensions of the cases. For instance, in terms of capability levels, we have two major-major, two major-minor, and six minor-minor power cases. Some of the actors have acquired nuclear weapons and some have not yet done so. As suggested above some actors rely or have relied on Soviet support, others on U.S. support, and still others on Chinese support. The rivalries are located in as many as five different regions or subregions. Normally, some or all of these differences might encourage considerable restraint in comparing the ten rivalries. Nonetheless, it is not clear, at least at this point in the analysis, that any of these overt differences are critical to the assessment of the theory. Should something emerge as a critical stumbling block in the subsequent comparison, one or more of the cases might have to be treated more gingerly. Yet there is no a priori reason to assume that this must be the case before any comparison or incomparability problems are encountered. To the contrary, should the theory prove relatively useful despite all of these different elements, its explanatory efficacy will have been buttressed considerably.

CHAPTER 3

The Egyptian-Israeli Rivalry, 1948–1970

The Egyptian-Israeli antagonism has played a central role in the Arab-Israeli conflict in the post–World War II era. It accounts for five wars (the Palestine War of 1948, the Sinai War of 1956, the Six Day War of 1967, the War of Attrition of 1969–1970, and the October War of 1973), numerous international crises, and most of the interstate casualties in the Arab-Israeli conflict. The rivalry emerged in 1948 during the Palestine War and began de-escalating after the October War in 1973, a process that continued in some respects through 1979 with the signing of the Egyptian-Israeli Friendship Treaty by Anwar Sadat and Menachem Begin. It has never terminated completely, but it has definitely de-escalated to a point at which war between the two states seems unlikely, other things remaining equal. The first twenty-two years of this rivalry (1948–1970) were dominated by little or no change in Egyptian or Israeli leadership expectations that would result in new strategies for de-escalation. In this chapter we focus on the failed attempts to de-escalate the rivalry in the absence of expectational change through 1969.

The Emergence of the Rivalry during the Palestine War of 1948

Although the Egyptian-Israeli rivalry emerged in the Palestine War of 1948, we will start our discussion mainly with the rise of Nasser after the Free Officers Revolution in Egypt in 1952.[1] Since our model is designed to understand rivalry de-escalation and termination, we are less concerned about the conditions that gave rise to the rivalry. However, we do want to acknowledge that the beginning of the rivalry in the Palestine War of 1948 is connected to

Egypt's rivalries with other Arab states such as Jordan and Iraq and their struggle for dominance over Palestinian territory.

The decision by King Faruq of Egypt to go to war in 1948 was a byproduct of two major foreign policy challenges: eliminating British colonial presence in the country and containing King Abdullah of Jordan from accruing more Arab territory, not just in Palestine but in Syria and Lebanon as well. These two issues would continue to influence Egypt's rivalry relationship with Israel throughout the Nasser years as well.

Between 1945 and 1952, Faruq's foreign policy was consumed with the central issue of eliminating the British political and military presence from Egypt. Faced with rising pressure from radical groups demanding Britain's withdrawal, Egypt's leaders knew that their public legitimacy depended on their ability to advance a nationalist agenda. Unfortunately, the British were not prepared to leave an area that they deemed to be vital to their strategic concerns. Yet, Egypt had little power to oust the British, who occupied the Canal Zone, Cairo, and Alexandria with fifty thousand troops (McNamara 2003: 16).

In 1945, King Faruq directed his efforts at undermining Britain's influence not just in Egypt but in the region as well through the creation of the Arab League, a pan-Arab organization headquartered in Cairo. Within the Arab League, two blocs of Arab states had emerged: the Hashemite powers of Iraq and Jordan, who supported Britain's presence in the region against the Triangle Alliance, composed of Egypt, Saudi Arabia, Syria, and Lebanon, who wanted to eliminate it. Despite the cleavages, Egypt dominated the Arab League and infused its policies with pan-Arab, anti-Zionist, and anti-imperialist ideologies—all of which resonated deeply among the Arab masses. In addition, Faruq embraced the idea of creating a defensive bloc of Arab states in the context of the Arab League that would be completely independent of Western states and whose members would be responsible to the will of the majority (Doran 1999: 76–80).

Not only did the Arab League rally around an effort to contain if not eliminate the plans of Britain and its Hashemite allies, it also unified the Arab countries around the commitment to Palestinian self-determination. The "Consensus Position" of the Arab League called for the creation of a Palestinian Arab state with sovereignty over all Palestine, rejecting Jewish and Arab cantons within a single state. The league condemned partition as a solution and endorsed a Palestinian state with the Arabs as a majority. This policy negatively affected Britain's ability to find an Arab leader who would mediate a solution between Arabs and Zionists. When it was unable to maintain control over Palestine as civil conflict between local Arabs and Jews escalated, Britain referred the Mandate of Palestine to the United Nations in February 1947 (Doran 1999: 100–101).

At that point, the Arab national movement in Palestine and the neighboring Arab states refused to recognize the partition plan, and launched a war instead. Egypt intervened in the Palestine War of 1948 not so much to eradicate the new fledgling Jewish state as to stop Britain and its Hashemite allies of Jordan and Iraq from accruing more Arab territory in Palestine and adversely affecting the Arab balance of power in the region (Doran 1999).

The key issue for Egypt (as well as the other Arab states) was the kind of political authority that would replace the British in Palestine. King Faruq, along with Syrian and Saudi leaders (as members of the Triangle Alliance), believed that the British would endorse an agreement to partition Palestine between Jordan and Israel. They perceived that in return for King Abdullah's recognition of Israel's existence, Israeli leaders might support his plan for a Greater Syria.[2] The evidence today indicates that for the most part they perceived correctly (Shlaim 2001b).

The Egyptians knew that the costs of nonintervention would have been too high. The Jordanians would have inevitably expanded into Arab Palestine and threatened to absorb Syria, or at best Syria would become someone's satellite. In either situation, the Triangle Alliance would no longer be a venue from which Egypt could dominate the Middle East. According to Doran (2001: 102), "alliance maintenance" dictated that Egypt enter the war so it could weaken Jordan and frustrate its plans for strengthening the British Empire. Moreover, Egypt was unwilling to leave the Palestine question to the exclusive machinations of Jordan, Israel, and the West (Doran 1999: 141).[3]

The Impact of the Palestine War of 1948: Israel Stonewalls

When the Palestine War was over, the Egyptians had suffered a major political and military defeat. Yet, during the course of the war, they were able to undermine Jordanian-Israeli cooperation, and the war left Jordan weak enough that it could no longer pose a threat to Syria. Nonetheless, the war also contributed to the destabilization of Syria, leading to three coups in 1949, and the Egyptians worried that Iraq and Iraq's supporters in Syria would seek a federation or failing that, that Israel would dominate Syria. Consequently, the Egyptians reaffirmed their commitments to pan-Arabism and their support for a regional defense system in the context of the Arab League. This Arab security pact was Egypt's ultimate grand strategy because its emphasis on Arab independence and pan-Arabism "tied together policies in all the major spheres of public life—the domestic, inter-Arab, Arab-Israeli, and great-power arenas" (Doran 2001: 103–104).

In 1949, Egypt signed an armistice agreement with Israel, regarding it as a temporary truce. Israel, in contrast, saw the armistice as a first step toward peace. By the 1950s, the Egyptian-Israeli conflict (like the general Arab-Israeli dispute) was centered on three issues: 1) Israel's return of border areas taken during the war of 1948; 2) the repatriation of over seven hundred thousand Palestinians to their homes; and 3) the internationalization of Jerusalem. While Israeli leaders were inflexible about making concessions on their new territorial gains and the fate of the Palestinian refugees, the Egyptians, like the Syrians and the Jordanians, were also unwilling to make concessions of their own, largely due to their fears about violent domestic repercussions should they sign any peace agreement with the Israelis (Oren 1992: 6).

Although publically the Egyptians appeared to be inflexible on these issues, privately they sought some accommodation with the Israelis. Between 1948 and 1952, the Egyptians and Israelis engaged in a series of secret contacts over the conditions of a final peace agreement.[4] King Faruq made several secret overtures to the Israelis for a permanent peace agreement in exchange for all or parts of the Negev Desert, retention of the Gaza Strip, and an accommodation for the Palestinian refugees (Oren 1992: 95–99). Consistently, the Israelis turned down the offers for any territorial concessions or repatriation. The fundamental reason for this intransigence was Prime Minister Ben-Gurion's belief that "time was on Israel's side": that with the passage of time, Israel's bargaining position would only improve; that the Arabs, the United Nations, and the Great Powers would eventually come to accept the status quo; and that the armistice agreement with Egypt was good enough to meet Israel's needs for "external recognition, security and stability" (Shlaim 2001a: 41–53).

For Egypt, its policies toward Israel would become more severely constrained as time passed. As anti-Zionism became synonymous with Arabism,

Egyptian leaders and domestic revolutionary parties alike set a high priority on Palestine's liberation. Since Israel's very existence reminded Arabs of their military debacle and humiliation in 1948, Arab leaders, especially those that aspired to regional dominance, felt compelled to pursue increasingly more aggressive positions against Israel on behalf of the Palestinians in order to legitimize their own governments and censure those that were considered too soft. Consequently, the rivalry between Egypt and Israel over the next two decades became more and more entangled with the political struggle for Arab leadership in the Middle East (Oren 1992: 7).

Internal Shock in Egypt: The Free Officers Revolution of 1952 and the Rise of Nasser

In July 1952 the Free Officers of the Egyptian army overthrew King Faruq and established a new regime with Major General Muhammad Naguib nominally in charge.[5] However, the Revolutionary Command Council (RCC), which was dominated by younger, more radical Free Officers such as Gamal Abdal Nasser and Anwar Sadat, established the policies for the new regime. Although there was no single political view among the Free Officers, they established two primary goals for the regime: the elimination of the British military presence in Egypt and the implementation of major political and economic reforms. In January 1953 the RCC dissolved all political parties (including its nemesis, the Wafd party) and declared Egypt a formal republic six months later. In the middle of 1953, Nasser, as the leader of the RCC, began to marginalize Naguib from key policy decisions. In November, Nasser consolidated his political control over the government when he accused Naguib of conspiring with the Muslim Brotherhood and removed him from the presidency. At that point, Nasser assumed the office and held it until his death in 1970.

At the outset of their rule in 1952 and 1953, Naguib and the RCC were less concerned about Palestine and Israel than they were about ending British influence in Egypt, instituting national reforms, and securing Egypt's leadership position in the Arab world. Consequently, they participated in secret communications with the Israelis primarily to maintain the status quo and avoid another round of conflict (Oren 1992: 101–103). In short, Egypt preferred to maintain the "no war, no peace" status quo (Morris 1993: 271).

On the Israeli side, Prime Minister Ben-Gurion held high hopes for a new beginning in Egyptian-Israeli relations. In a major speech before the Knesset in August 1952, he maintained that there was no reason for political or military conflict between the two countries, and he followed his speech with a major peace initiative. However, Naguib and the RCC did not respond and Ben-Gurion suspected that the Egyptians were projecting an image of moderation that would enhance Egypt's ability to get arms and aid without first taking serious steps toward a settlement with Israel (Shlaim 2001a: 78).

However, a new development occurred in the first half of 1953 when Nasser took charge of the RCC's negotiations with Israel. Nasser sent a message indicating that Egypt was not departing from its pan-Arab policy on Palestine. However, he requested Israel's support in obtaining economic aid for Egypt from the United States and for its support in Egypt's efforts to terminate the British military presence in the Suez Canal Zone. Nasser also made it clear that he wanted these negotiations to remain secret (Shlaim 2001a: 79).

In response to Nasser's initiative, the Israelis extended a serious offer at reconciling the differences between the two countries. They were willing to provide economic assistance to Egypt by purchasing $5 million worth of cotton and other products if Egypt lifted its restrictions on the passage of Israeli oil tankers through the Suez Canal and the Gulf of Aqaba. Next, the Israelis would be willing to advance their support for British withdrawal from the Suez Canal if the Egyptians first showed their commitment to pursuing better relations via a secret high-level meeting. Eventually, Nasser turned down Israel's offer to buy Egyptian cotton, but the RCC did look into the matter of easing restrictions for Israeli ships through the Suez Canal (Shlaim 2001a: 79–80).

Ben-Gurion reacted pessimistically. He believed that the Egyptians were asking for too much and giving too little in return. He sent a counterresponse that stated Israel would be willing to help with the Suez issue if Egypt allowed free passage of Israeli ships through the Suez and Eilat as well as commit to a secret high-level meeting, which was a crucial test of Nasser's intentions. The Egyptians made no formal reply, and later Israel would learn through its Egyptian contacts that Nasser had decided not to pursue a request for Israel's help in securing U.S. economic aid because he did not want to incur an obligation. He also decided to bypass the secret meeting until the Suez dispute with the British was resolved. Afterward, the Israelis perceived that the RCC and Nasser had not been serious about pursuing peace negotiations, and by the end of 1953, Israel's high hopes for better relations had begun to disappear (Shlaim 2001a: 81).

Two External Shocks Intensify the Egyptian-Israeli Rivalry

Gaza Raid of February 28, 1955

In 1955, two foreign policy shocks significantly altered the relations between Egypt and Israel. The first shock emerged from increasing cross-border infiltrations from the Gaza Strip. By 1955, the frequency of these border incursions was posing serious day-to-day security problems for the Israelis, who were never really sure about treating them as isolated events or interpreting them as indicators of a more hard-line Egyptian foreign policy. The second shock occurred when Egypt signed a Soviet-backed arms deal with Czechoslovakia worth $320 million. Israeli leaders perceived that the deal altered the military balance of power, and they were convinced that Nasser was getting ready for a second round of war (Tal 1996).

Before the Egyptian-Czech arms deal, the Israelis were more concerned with upholding the status quo and protecting Israeli civilians and property along the border areas controlled by Egypt. In an effort to compel the Egyptians to take stronger preventive measures that would stop the infiltrations, the Israelis settled on a policy of reprisals. Generally, the lion's share of the border incursions from Gaza involved Palestinian refugees who traveled into Israel to visit relatives, seek food, harvest crops, or steal from Israeli work projects. A smaller number of these crossings involved Palestinian activists who engaged in sabotage and murder for the purpose of challenging the armistice lines and keeping the general conflict with Israel alive. Eventually, Egyptian intelligence agents and some soldiers participated in these crossings as well. In response, the Israelis pursued a reprisal strategy that shifted from initially killing individuals caught in Israel, to raiding Arab villages in Gaza, and in 1954 attacking Egyptian military forces both in Gaza and the Al-Auja demilitarized zone.[6] Over time, the escalation of border crossings and reprisals culminated in a very controversial attack by Israel against Egyptians troops in Gaza on February 28, 1955. In that attack, Israeli troops killed thirty-eight Egyptian soldiers (Morris 1993; Tal 1996).

The Gaza raid was a major turning point in Egyptian-Israeli relations that ended the possibilities of any settlement.[7] The scale and damage of the operation both shocked and humiliated Nasser and his armed forces. Until this time, Nasser had maintained a consistent and firm policy of curbing Palestinian infiltration from the Gaza Strip in an effort to keep the border areas quiet and maintain the status quo (Morris 1993; Shlaim 2001a). After the raid, Palestinian refugees engaged in three days of violent demonstrations and riots against Egyptian and UN buildings and jeopardized Nasser's control over Gaza. Not only did the raid endanger Nasser's domestic political position, it also altered Nasser's perceptions of the Israelis. He no longer believed in the possibility of coexistence with Israel and he felt personally responsible for the Egyptian casualties. He also maintained that the raid had hurt his regime's prestige, and his military government could not suffer such a defeat again without retaliation (Morris 1993: 328–329).

At this point, Nasser changed his policy of curbing and repressing cross-border infiltration to one that would allow the refugees to engage in militant actions that fell short of war. Now, he supported retaliatory strikes against Israel and he unleashed a series of raids and counterraids against Israeli targets. More importantly, Egypt embarked on a policy of recruiting and organizing fedayeen units in Jordan, Lebanon, and Syria to raid Israeli targets from third countries (Shlaim 2001a: 125–127).

In August and September 1955, the border clashes between Egypt and Israel peaked with Israel's attack in Khan Yunis in Gaza, which killed seventy two Egyptians and Palestinians. These heavy casualties coupled with pressure from the United States convinced Nasser to back off and pursue different options:[8] tighter restrictions in the Straits of Tiran against Israeli shipping, the closure of air space over the Gulf of Aqaba to Israeli aircraft, and support for more third-party infiltrations by fedayeen in other Arab countries (Morris 1993: 327–350).

On the Israeli side, the Gaza raid in February 1955 signaled a shift in Israeli policy. After returning to the Defense Ministry in 1955 (after a two-year hiatus from the government), Ben-Gurion believed that Nasser was an "implacable and dangerous enemy" and argued that Prime Minister Sharett's conciliatory policies had to end (Shlaim 2001a: 123).[9] Within days of his return to office, Ben-Gurion inaugurated a tougher approach by approving the Gaza raid in February. He worried that Nasser's ascendancy in the Arab world represented a serious threat to Israeli security. Therefore, he hoped to use a military operation to weaken Nasser by exposing his military's weaknesses. He argued that it was important to display Israel's military superiority over the strongest Arab country as well as bolster the confidence of the Israeli public and army (Shlaim 2001a: 125). According to Tal (1996: 64), Israel's decision to launch the Gaza raid also occurred against a backdrop of important events that undermined Israel's sense of security, particularly Egypt's success in negotiating the withdrawal of British troops from the Suez Canal in 1954.[10] Israelis also feared that the United States would provide economic and military assistance to Egypt in its efforts to entice Nasser to join a regional defense system and that Egypt would use the arms against Israel.

The Egyptian-Czech Arms Deal of 1955

Another major shock to Egyptian-Israeli relations was the $320 million Soviet-backed arms deal between Egypt and Czechoslovakia.[11] The Gaza raid played a significant role in Nasser's decision to negotiate an arms deal in May 1955, with the first deliveries reaching Egypt in July (McNamara 2003; Morris 1993; Shimshoni 1988). Since the raid demonstrated Egyptian military inferiority vis-à-vis the Israeli Defense Forces, Nasser searched more aggressively for military arms to equalize the situation.[12] When the United States and Britain refused to comply, Nasser turned to the Eastern Bloc.[13] Although the Gaza raid was an important factor, other scholars suggest that it was more a pretext than a cause for the arms deal (Bar-On 1994; Shlaim 2001a; Doran 2001). Actually, Egypt's arms deal had to be considered in a wider inter-Arab context; namely, the Gaza raid also coincided with the formal signing of the Baghdad Pact between Turkey and Iraq in February 1955 with Iran, Pakistan, Great Britain, and hopefully Jordan joining soon after.[14]

The Baghdad Pact threatened to undermine Egypt's long-term efforts to eradicate Britain's regional security system, and it damaged Nasser's efforts to secure a leadership role in the Arab region.[15] He believed that the United States and Great Britain would expand the regional role of Jordan and Iraq, and even bring about the Greater Syria project, at the expense of Egypt and the Arab League. In short, the Baghdad Pact renewed the competition between the Egyptian-led Triangle Alliance and the pro-British Hashemite states for Arab leadership. Nasser secured the Egyptian-Czech arms deal in an attempt to establish greater independence from Britain, destroy the

Baghdad Pact, and ensure the eradication of Britain's military presence from the region. Rhetorically, Nasser explained to his constituents both at home and in the region that the Baghdad Pact represented "perpetual subservience to Zionism and imperialism," while the Arab League led by Egypt would be faithful to the priorities and values of the Arab world (Doran 2001: 108). Hence, the attempts by the Great Powers to build an anti-Soviet regional defense system encouraged Nasser to be more aggressive in asserting his pan-Arab leadership, which produced a more hard-line policy vis-à-vis Israel.[16]

For Israel, the Egyptian-Czech arms deal was a "watershed" development in Israeli-Egyptian relations because it immediately changed the prevailing balance of power. Israeli military planners estimated that Egypt's arsenal would now be three times larger than what Israel had or expected to get in the short term. However, they were more concerned about the qualitative shift in the military balance of power. The newer sophisticated Soviet weapons challenged the Israelis' view that their technological superiority could overcome any numerical advantage that the Egyptians had. Since these weapons deprived the Israelis of a "deterrent effect," Nasser was likely to exploit the situation and attack Israel as soon as his army had absorbed the weapons. Ben-Gurion and his military planners believed that the Egyptians were likely to attack as early as the summer of 1956 (Bar-On 1994: 17–18; Morris 1993: 277).[17]

Arms Buildup and the Sinai War in 1956

When Britain and the United States refused to abrogate the Tripartite Agreement of 1949,[18] the Israelis turned to France, which agreed to sell arms, primarily because of French antagonism toward Nasser, who was sending arms to Algerian nationalists. The French hoped that if Israel could eliminate Nasser, the Algerian rebellion would collapse (Shlaim 2001a: 163). In July 1956 the Israelis signed a contract with the French to purchase $51 million of French arms, including 36 jets, 210 tanks, and 100 guns (Oren 1992: 91). After the Israelis contracted for arms with the French, Nasser successfully negotiated a second round of weapons with the Soviets for more aircraft jets, much to the dismay of the United States.

Meanwhile, the United States tried to mediate a comprehensive peace settlement between Israel and the Arabs (for example, the Alpha Project), but it failed to produce any tangible results. The final breach of good relations between Egypt and the United States and Britain occurred at the end of 1955 when Jordan became the battleground over the Baghdad Pact. After King Hussein showed a willingness to join the pact, Nasser and the Saudis generated widespread demonstrations in Jordan that prevented Hussein from joining. At this point, the United States, believing that Nasser was opposed to its interests in the area, withdrew its funding for the Aswan Dam on July 19, 1956. On July 26, Nasser nationalized the British- and French-owned Suez Canal. During the same time period (the summer of 1956), the Israelis began to prepare for a "preventive war" and through their talks with the British and the French signed onto to a joint military operation that would initiate the Suez War on October 29, 1956 (McNamara 2003: 53–59).

Within days, the Israelis achieved a complete military victory. The Egyptian forces withdrew from the Sinai and the Gaza Strip, leaving six thousand soldiers behind, which minimized their losses to the army. By November 2 the Israelis had captured Gaza and three days later all of the Sinai was in their hands. During the war, Eisenhower pressured the British, the French, and the Israelis to cease their military operations.[19] In particular, he demanded an unconditional Israeli withdrawal and threatened to cease all U.S. government and private aid to Israel if it failed to do so. However, four months later, the United States negotiated a deal that would ensure Israel's right to navigate the Straits of Tiran and the Red Sea. The United States went so far as to say that if Egypt renewed its blockade, Israel had the right to use military force in the interest of "self-defense" (Shlaim 2001a: 181–183).[20]

The Impact of the Sinai War (1956–1966): Egypt and Israel Continue Hard-line Policies

According to Shlaim (2001a: 183), the Israelis had three military objectives—all of which they achieved—and three political ones—none of which were completely successful. On the military front, the Israelis defeated the Egyptian army quickly, and their victory reestablished their superiority and their deterrent capability in the Middle East. Regarding the international waterways at the Straits of Tiran, U.S. guarantees assured that Israeli shipping would not be curtailed. Since Israel had destroyed the fedayeen bases in Gaza, fedayeen attacks across the border ceased. Moreover, the Egyptian forces did not return to their bases in the Sinai. In that respect, the Sinai became "demilitarized." Israel was able to experience eleven years of relative border peace and stability with Egypt.

Nonetheless, on its three political objectives, the results were mixed. The Sinai War did not result in the overthrow of Nasser; instead, Nasser emerged with even more prestige and influence in the Middle East and the periphery. Next, the Israelis were not able to extend their borders. Nonetheless, the war gave Israel "unqualified sovereignty" within these borders and the quarrel with Egypt over the interpretation of the armistice agreement of 1948 was settled in Israel's favor (Bar-On 1994: 319). Finally, Ben-Gurion had hoped that war would bring down or weaken the governments in Lebanon, Iraq, and Egypt to such an extent that each of them would make peace with Israel on its terms. Instead, the war of 1956 solidified pan-Arab commitments to resolving the Palestine problem while the countries in question worked on changing the military balance of power with Israel (Shlaim 2001a: 183–185).

The war did convince Ben-Gurion to accept the territorial status quo outlined in the armistice agreement of 1949. Adopting a strategy of deterrence that would discourage Egypt and its allies from changing the territorial status quo through military force, he upgraded Israel's military capabilities significantly to ensure a qualitative advantage over his Arab rivals. He also sought international diplomatic guarantees for Israel's security. He was especially worried about the Soviets extending their influence in the region through their military and economic support for radical Arab regimes that were most hostile to Israel's security (Shlaim 2001a: 189).

As for Nasser, he emerged from the war without a clear strategy vis-à-vis Israel. Instead, he opted for a policy of hostility short of war: economic boycotts, strategic blockade (via the Suez Canal), sporadic Egyptian-supported guerrilla attacks by Palestinians, political warfare in the international community, and continued pressure on Israel to allow Palestinian refugees to return home. In short, he confined the Palestinian problem to international diplomacy and he avoided situations that increased the risks of war with Israel (Sela 1998: 52–57).

Throughout the late 1950s and until May 1967, Nasser repeatedly argued that the war against Israel needed to be postponed in order to give Egyptians and other Arabs the time to build up their military, economic, and political capabilities (Sela 1998: 52). Instead, he preferred to concentrate on consolidating his socialist economic revolution at home and creating a unified Arab regional hegemony through his support for pro-Nasserist parties and governments throughout the Middle East. At home, Nasser initiated a state-led import-substitution industrialization policy that was financed by nationalizing the nonagricultural private sector of the economy and relying heavily on external aid from abroad.[21]

On the regional front, Nasser was busy extending his vision of pan-Arab unity throughout the Arab Middle East. His regional efforts culminated in 1958 with the Egyptian union with Syria, whose regime was also a socialist government closely aligned with the Soviet Union. The unification—resulting in the establishment of the United Arab Republic—was a high point for Nasser's regional foreign policy. However, by 1961, a military coup in Syria had ended the union, and Nasser embarked on a more radical regional policy that supported internal revolts against conservative regimes in Jordan, Iraq, Lebanon, and Saudi Arabia. In October 1962, Egypt intervened in the Yemeni civil war, which rapidly escalated to a protracted five-year proxy war between Egypt and Saudi Arabia. At this point, Nasser had little interest in confronting Israel.

Israeli-Syrian Clashes and the Escalation to the Six Day War in 1967

Ironically, the Egyptian front with Israel was peaceful for a full decade after the Sinai War, primarily as a result of Nasser's military commitments in Yemen and his preoccupation with inter-Arab struggles. The most serious border conflicts in the 1960s occurred between Israel and Syria over the Golan Heights. Two particular issues were at stake: control over water resources from the Jordan River within the demilitarized zones between Israel and Syria, and military attacks from Palestinian guerrilla organizations (Shlaim 2001a: 228).

Throughout the 1950s and in the early 1960s the Israelis engaged in provocative military raids against the Syrians in an effort to capture areas within the demilitarized zones for the purpose of diverting the Jordan River for a major irrigation project in the Negev and gaining control over the Golan Heights in order to enhance their security against Arab infiltrations and shellings of Israeli settlements (Slater 2002: 90; Shlaim 2001a: 235–236).[22] According to Shlaim (2001a: 235), Israel's clash with the Syrians was "probably the single most important factor in dragging the Middle East to war in June 1967."

In early February 1966, an intraparty military coup produced a more radical regime in Syria. The new Ba'thist government, frustrated with the lack of containing Israeli aggression through the UN Security Council, decided to take stronger retaliatory actions against Israel. As the Syrians escalated their clashes with the Israelis, they also encouraged Palestinian guerrilla attacks against Israeli targets. Between February 1966 and May 1967, Syria initiated 177 border incidents and 75 Palestinian guerrilla raids inside Israel (Ma'oz 1995: 89–93).

The escalating hostilities between Israel and Syria increased Soviet and Egyptian fears that a war with Israel could harm the new regime at the same time that it could drag Nasser into a war that he wanted to avoid. Consequently, the Soviets negotiated a rapprochement between Egypt and Syria that ended in the signing of a mutual defense pact in November 1966. For Nasser, the defense pact was designed to diminish Syria's insecurities about Israeli aggression and provide him with greater leverage to diminish Syria's use of force against Israel. Nonetheless, the Syrians refused to back down. They continued to support Palestinian guerrilla actions against Israel, rejected Egypt's offer to deploy its air force units on Syrian soil for added protection, and demanded that Nasser remove the UN Emergency Force (UNEF) troops from the Sinai (Sela 1998: 89–90).

In early 1967, the Israelis resumed their activity in the demilitarized zone, provoking clashes with the Syrians, and on April 7 the Israeli air force shot down six Soviet-made Syrian MiGs but not before penetrating Syria and flying over Damascus, humiliating the Syrian military. On May 12, 1967 Israel's defense minister, Yitzhak Rabin, threatened to overthrow the Syrian regime in a news interview. Although the Israelis quickly repudiated Rabin's comments, the Soviets became concerned and intervened in the crisis by sending a report to Nasser that Israel was mobilizing its forces on its northern border and planning to attack Syria. Although the report was false and Nasser knew it, he mobilized his troops into the Sinai on May 14, 1967. Next, he asked the United Nations to remove the UNEF troops from the Sinai, and on May 22, 1967, he closed the Straits of Tiran to Israeli shipping and expressed his willingness to go to war with Israel (Sela 1998: 87–90; Shlaim 2001a: 236–237; Feiler 2003: 196). Nasser believed his military's assessment that Egypt could withstand an initial military attack from Israel and launch a stronger counteroffensive that would significantly weaken Israel. He also believed that fighting between Egypt and Israel would end early due to the inevitable intervention from the United States and the Soviet Union, thus preserving Egypt's position in the straits (Feiler 2003: 203–204).

Although Nasser reiterated frequently that war with Israel should be postponed until the Arabs were stronger, he embarked on a "defensive strategy" while hoping that the United States or the Soviet Union would intervene to halt the conflict. Scholars maintain that Nasser's decision to risk a war with Israel is best explained by his declining leadership in the Arab world. After 1964, Nasser experienced heavy criticism from his Arab rivals about his unwillingness to confront Israel aggressively and his cowardice of hiding behind UN forces in the Sinai. Moreover, Nasser's defense commitment to Syria added more pressure on him to preserve his prestige and credibility. When Nasser mobilized his forces in the Sinai, Arab public opinion demanded even more militancy, and he complied by removing the UNEF troops from the border. Finally, his decision to blockade the Straits of Tiran constituted a "casus belli" for the Israelis. Two weeks later, the Israeli government launched a massive air and ground attack against Egypt, Jordan, and Syria, and within days Israel had defeated all of their military forces (Sela 1998: 92–93). At the end of the war, Israel had more than doubled its size, capturing the Sinai Peninsula, the Golan Heights, and the Jordanian-held West Bank, including all of Jerusalem.

The Impact of the Six Day War of 1967: No Overall Change in Egyptian-Israeli Relations

In Egypt's case, the military defeat in 1967 produced within the Egyptian national leadership two strategic alternatives to Israel. One approach was to pursue a political solution with the aid of the United States, which would be inclined to intervene in order to check the Soviets' increasing influence in the Middle East. The other approach was based on the presumption that a political solution was pointless and that only a buildup of military power would lead to a solution; the buildup would not necessarily be for war, but would serve as a means of convincing the Israelis that a just solution would have to be based on a balance of interests and on the principle that "what was taken by force would be returned by force" (Shemesh 2008: 353). Nasser opted mostly for the second solution but not without discarding a political approach altogether. Eventually, he would pursue a military option that would be complemented with a political approach.

At the inter-Arab conference in Khartoum in September 1967, Nasser was able to convince the Arab states to support a two-staged solution to the Arab-Israeli conflict. The first stage would involve the return of Arab territories that Israel occupied after the Six Day War, and the second stage would involve the goal to achieve the liberation of Palestinian lands and the creation of a Palestinian government. The summit meeting in Khartoum represented a "watershed" in terms of securing a solution in stages and in terms of pursuing a political path. The view held by Nasser and other Arab leaders was that it would take a long time before the first stage could be achieved by military means. Hence, they were more flexible about pursuing a diplomatic path (Shemesh 2008: 243).[23] Nasser even encouraged King Hussein to explore the possibility of negotiating a return of the West Bank and East Jerusalem to Jordan, although the Israelis did not know this at the time (Shlaim 2001a: 259).

More evidence of Nasser's willingness to pursue a political option was his agreement to accept the UN Security Council's Resolution 242 in November 1967.[24] In Nasser's view, this agreement laid the foundation for implementing the first stage of the Arab/Egyptian strategy conceived at the summit in Khartoum. Nasser's support of the UN resolution also led to the UN-sponsored Jarring mission from 1967 to 1969.[25] Such efforts also increased a dialogue between Egypt and the United States, which Nasser hoped would lead to U.S. pressure on Israel to withdraw its forces from newly gained Arab territory (Sela 1998: 108; Shemesh 2008: 243).

Despite these diplomatic overtures, Nasser also maintained a pragmatic but hard-line policy toward Israel, especially in light of growing domestic unrest over Egypt's military defeat, the deterioration of the economy between 1967 and 1970, and increased government repression and authoritarianism. In the midst of this internal turmoil, Nasser decided that recovering the Sinai had the highest priority, and he also believed that the only means to secure it was through another round of war with Israel. He perceived that war would not only redeem his declining prestige in the region but also shore up his political legitimacy at home. Hence, he refused to make any deals with Israel that entailed an exchange of the Sinai for a final peace agreement. Instead, he opted for a policy of rebuilding his armed forces but not without sending a clear message that he was open to negotiations that would result in the unqualified return of the Sinai even if it came at the expense of the Palestinians (Sela 1998: 97–99). But, for the short term, Nasser was interested in rebuilding his army and searching for a massive infusion of economic assistance to overcome the financial shortfalls associated with the war: closure of the Suez Canal, the decline in tourism, and the loss of the Sinai oil fields. It is estimated that the Egyptian economy lost $219–$230 million just with the closing of the Suez Canal alone (Feiler 2003: 174).[26]

The most significant impact that the Six Day War had on Nasser's foreign policy was the shift in his Arab regional policy. While at the conference in Khartoum, Nasser made peace with the oil-rich, conservative Arab governments by eschewing his pan-Arab radicalism in favor of championing an anti-Zionist movement in the Arab world. He argued successfully that as a front-line confrontation state against Israel Egypt needed significant amounts of financial aid to prepare it for the next round of warfare. In exchange for his withdrawal of Egyptian troops from Yemen,[27] his moderation of his revolutionary ideology, and his decision to move away from a strict socialist control over the Egyptian economy, Kuwait, Libya, and Saudi Arabia provided $257 million a year indefinitely—a figure that was 10 percent higher than Egypt's estimated losses (Feiler 2003: 7).[28] This Khartoum aid to Egypt continued even after the opening of the Suez Canal in 1975 and lasted until March 1979, when an economic boycott was levied against Egypt in retaliation for Egypt's participation in the Camp David Accords in 1978. In addition to funds from the oil-rich Arab states, Nasser also obtained more arms from the Soviets, who expanded their support from $424 million in weapons from 1963 to 1966 to $2.23 billion from 1967 to 1973 (Barnett 1992: 114–115). With a new infusion of foreign assistance, Nasser prepared for war since he perceived that the Israelis were unlikely to give up the Sinai voluntarily.

As for the Israelis, they debated the merits of a peace offer to the Egyptians in exchange for the demilitarization of the Sinai and security guarantees for Israeli rights of passage through the Suez Canal, the Straits of Tiran, and the Gulf of Aqaba.[29] After the Khartoum conference, however, the Israelis canceled the offer when the Arabs concluded their meeting with the famous three nays: no peace, no recognition, and no negotiations with Israel (Shlaim 2001a: 258). Although moderate Arab leaders won the day at the Arab summit by calling for a political solution rather than military means to seek Israeli withdrawal, the Israelis saw only a hard-line policy. At this point, they decided that Israel would maintain the status quo as determined by the ceasefire agreements and reinforce its position according to its security and development needs. More ominously, Israeli leaders secretly canceled the principle of seeking peace with Egypt on the basis of the international border in favor of one based on secure borders (Shlaim 2001a: 259).

But, the three nays at the Khartoum conference were not the only reason Israel decided to pursue a hard-line policy. By January 1968, a new coalition of political parties, referred to as the second labor alignment, was dominating the government under Prime Minister Levi Eshkol. The new Labor Party incorporated a wide range of views on Israel's foreign policy—some of which were committed to territorial compromise, some of which wanted to preserve Israel's hold on all the territories or simply expand its control over the West Bank. The wide diversity of views made it difficult for the Israeli government to pursue negotiations seriously without alienating a key faction and bringing down the government. Hence, the lack of consensus within the governing coalition contributed to the government's immobilization on foreign policy initiatives (Shlaim 2001a: 262). After Golda Meir became prime minister in March 1969, the hawks, who were opposed to any territorial compromise, disproportionately dominated her cabinet, and Israeli foreign policy consisted of "military activism and diplomatic immobility" (Shlaim 2001a: 289). Meir preferred the status quo over domestic political costs that might be incurred through diplomacy and compromise. Therefore, she refused to shift from the ceasefire lines of 1967 until the Arabs accepted a peace agreement on Israel's terms. According to Shlaim (2001a: 289), "intransigence" was not only Meir's middle name; it was also the hallmark of Israel's policy toward the Arabs for the next five years.

The Impact of the War of Attrition, 1969–1970: No Change in Rivalry Relations

After the Six Day War, Nasser moved ahead with his plans to rebuild his armed forces. He concentrated on reorganizing the army, adding three divisions to his ground forces, centralizing its leadership, and retraining and reeducating his soldiers. Moreover, he invested tremendous resources in his air defense and air force primarily through Soviet aid.[30] By early 1969, Egypt had restored its prewar strength and expanded its combat aircraft. Although it had a greater number of planes than Israel, its pilots were not as skilled as the Israelis, which ruled out a direct frontal assault against Israel. Instead, Nasser opted for a limited war against Israel along the Suez Canal area with the aim of inflicting large numbers of casualties over time. Nasser's goal was to prevent the solidifying of the military status quo after the Six Day War and to pressure the international community to force Israel to withdraw from the Sinai Peninsula and other occupied Arab territories (Yaqub 2007: 38).

In return, the Israelis settled on a "defensive strategy" of preventing the Egyptian army from crossing the Suez Canal and capturing territory on the east bank. Hence, they built a line of fortifications along the canal (for example, the Bar Lev Line) where they concentrated their military forces. At the beginning of 1970, the Israelis shifted to a strategic bombing campaign against Egypt's inner cities. The Israelis anticipated that the bombing would bring the war to an end and compel Egypt to recognize the ceasefire. On a political level, they expected that the deep penetration bombing would break Egyptian morale, bring about Nasser's political demise, and produce a pro-Western successor. Between January and April 1970, Israel flew thirty-three hundred sorties and dropped an estimated eight thousand tons of ordnance on Egyptian territory (Shlaim 2001a: 291–294).[31]

When Nasser embarked on what became the War of Attrition, he failed to anticipate that Israel would counter with strategic bombing campaigns deep in Egyptian territory. The costs were in some cases higher than the Six Day War in 1967. For instance, the death toll exceeded the last war, and Israel's destruction of the Suez Canal cities produced a massive population exodus to Cairo, where many of the displaced became unemployed and impoverished. The military stalemate in the war combined with increasing economic burdens influenced Nasser to appeal to the United States to intervene in the conflict and force Israel to withdraw from the occupied Arab territories. In July 1970, Nasser accepted the U.S.-backed Rogers Plan for a ceasefire of hostilities along the Suez Canal military zones and the resumption of a UN peace mission. Shortly afterward, the Israelis also agreed to the plan (Feiler 2003: 177; Yaqub 2007: 42).

The State of the Egyptian Economy by the End of the 1960s

By the time of Nasser's death in September 1970, Nasser's Five Year Economic Plan in 1960 had failed to generate economic development and growth as it was intended to do. By the mid-1960s, the Egyptian economy was characterized by several ominous trends: a) increasing public consumption at the expense of domestic investment; b) rising trade imbalances due to poor export performance; c) growing defense outlays associated with war against Israel that constituted as much as 20 percent of Egypt's GDP in 1973; and d) a series of external shocks that undermined economic growth—for example, the failure of the cotton crop in 1961, the costs of the Yemeni civil war (1962–1970), and suspension of U.S. food imports (Kenawy 2009: 591; Waterbury 1983: 91–93). As a consequence, Nasser had to rely more frequently on external debt financing especially after the Six Day War of 1967.

Although aid from the Soviet Union increased, it was unable to compensate for the reduction in Western aid. The main source of external aid after 1967 came from the Arab oil countries as a result of the Khartoum agreement in 1968. Saudi Arabia, Kuwait, and Libya provided $286 million on average annually. Although this aid was significant, Nasser faced another set of economic burdens, which included newly emerging payment deadlines for loans taken for ambitious development projects before 1967. These payment installments amounted to $240 million annually from 1967 to 1972. Refusing to undertake more economic reforms that would reduce public and private consumption, Nasser continued to pursue policies that would regain the Sinai from Israel and restore Egypt's lost prestige. This strategy came at the expense of future economic growth and growing foreign indebtedness (Waterbury 1983: 98–99). In short, at the time of Nasser's death in September 1970, the Egyptian economy was on the verge of bankruptcy.

The Application of the Expectancy Revision Model to the Egyptian-Israeli Rivalry of 1948–1970

Table 3.1 outlines the key variables of our model. It shows that while Nasser had consolidated his political control over Egyptian institutions as president during the years 1950–1970, he remained inflexible in the presence of internal and external shocks and the absence of third-party pressure, reciprocity, and/or reinforcement toward de-escalation from Israel.

On the Israeli side, the years 1950–1970 are associated with four Israeli prime ministers—only one of whom was clearly in charge of Israel's foreign policy without any serious domestic constraints and that was David Ben-Gurion, who monopolized the policymaking in determining Israel's security strategies and national priorities from 1948 to 1953. Although his reign

Table 3.1. Modeling Outcomes for the Egyptian-Israeli Rivalry, 1948–1970

Variable	*Egyptian president Nasser 1954–1970*	*Israeli prime minister Ben-Gurion 1948–1963*	*Israeli prime minister Sharett 1954–1955*	*Israeli prime minister Eshkol 1964–1968*	*Israeli prime minister Meir 1969–1970*
Shocks	Present	Present	Present	Present	Present
Expectancy Revision	Absent	Absent	Absent	Absent	Absent
Consolidated and new policy entrepreneur	Absent	Absent	Absent	Absent	Absent
Third-party pressure	Absent	Absent	Absent	Absent	Absent
Reciprocity	Absent	Absent	Absent	Absent	Absent
Reinforcement	Absent	Absent	Absent	Absent	Absent
Outcome	No change	No change	No change	No change	No change

Note: Ben-Gurion served 1948–1953 and 1956–1963; Meir served 1969–1974.

as prime minister was interrupted for two years (1954–1955), Ben-Gurion was able to reestablish his control over Israel's foreign and defense policies largely as a result of Israel's successful involvement in the Sinai War of 1956. He was an indisputable leader because he was able to win majorities in the Knesset and his cabinet at will (Shlaim 2001a: 187). He also dominated the Israeli Defense Forces by appointing a military chief-of-staff, Major General Chaim Laskov, in 1958; Laskov stayed out of politics and deferred to Ben-Gurion's policies. Hence, in the years 1958–1963, Ben-Gurion again enjoyed a near monopoly over Israel's foreign and defense policies (Shlaim 2001a: 187–188). Nevertheless, this policy control never corresponded with a sea change in Ben-Gurion's views of Egypt or those of his successor. Table 4.1 in Chapter 4 also indicates that while none of the four Israeli prime ministers in the years 1950–1970 had changed their expectations about their Egyptian counterpart, there was also an absence of reciprocity and reinforcement from the Egyptians and a lack of international third-party pressure despite the presence of major internal and external shocks during the regimes in question.

Between 1952 and 1970, the shocks, outlined in Table 3.2, emanated from the external, regional environment rather than from major developments internal to either Egypt or Israel. Unfortunately, they reinforced

Table 3.2. Shocks in the Egyptian-Israeli Rivalry, 1948–1970

Time frame of shocks	*Shift in external threat*	*Shift in competitive ability*	*Shift in regime orientations/ strategies*	*Economic resource crisis*
1948–1949		Egyptian war defeat		
1955		Gaza raid; Czech arms deal		
1967		Egyptian war defeat		
1969–1970		Egyptian war defeat		

rather than changed preexisting and future expectations of threat on both the Egyptian and Israeli sides. The Israeli raids into Gaza in 1955, the Sinai War in 1956, the Six Day War in 1967, and the subsequent War of Attrition in 1969–1970 only hardened Nasser's views toward the Israelis and vice versa. The severe deterioration of the Egyptian economy might have had an inhibitive effect in other circumstances, but just enough foreign assistance was forthcoming to allow decision makers to evade its policy implications. Egypt's race to build up its military capabilities, its support for Palestinian covert operations against Israeli targets, and its bellicosity prior to the Six Day War in 1967 created a perceived opportunity for Israel to win another war and hopefully dismantle Nasser's regime.

The expectancy model also stresses the key role that policy entrepreneurs play in overturning preexisting policy monopolies held by opposing factions inside and outside their governments. Policy entrepreneurs are able to bring about fundamental shifts in their foreign policies because they are well positioned to push new thinking and policies through these opposing forces. In short, policy entrepreneurs have the necessary support of their domestic constituencies to support such changes. However, the presence of policy entrepreneurs does not always correlate with dramatic changes regardless of their political influence and their control over their domestic constituencies or bureaucracies. For instance, Nasser had consolidated his political control over the Egyptian government by 1953 and he enjoyed widespread public support. However, this position did not translate into new directions in Egyptian foreign policy, and despite the numerous shocks that adversely affected Egypt both regionally and domestically, Nasser refused to move in fundamentally new directions. As for the Israeli prime ministers between 1952 and 1970, only Ben-Gurion was in a position to monopolize foreign policy and he also refused to change his thinking despite continued reinforcement that hostilities between Israel and Egypt were not de-escalating in the face of Egypt's military defeats.

The key actors in this period were the United States and the Soviet Union, but neither of them took an overt diplomatic position to resolving the conflict. Their Cold War hostilities played out in the set of alliances that the United States and the Soviet Union established in the Middle East, which overlapped with the Arab-Israeli conflict. For Nasser, U.S. alliances with his Arab and Israeli competitors blocked any potential efforts that the United States could have undertaken to reduce the rivalry. Moreover, the United States had little incentive to pressure the Israelis to change their policies in light of Egypt's close ties to the Soviet Union and its fear of growing Soviet influence in the area. The Soviets in turn extended foreign aid and military arms to its allies, particularly Egypt and Syria, who clashed repeatedly with Israel. Besides intermittent efforts by the United Nations, there was little sustained diplomatic effort by the superpowers to negotiate a rapprochement between Egypt and Israel in light of the Cold War cleavages in the region.

There was one overture that had the potential to lead to more cooperation, but it failed to be pursued by either Egypt or Israel. During his early years, Nasser explored the possibility of cooperation with Israel for securing economic aid from the United States and influencing the British to leave the Suez Canal in 1953. However, Nasser failed to reinforce his offer of cooperation either publically or by offering to reduce restrictions for Israeli ships going through the Suez Canal. Consequently, the Israeli government under Ben-Gurion did not pursue any more diplomatic efforts. This state of affairs continued until 1970, when Anwar Sadat of Egypt undertook a significant peace initiative—the focus of the next chapter.

CHAPTER 4

The Egyptian-Israeli Rivalry, 1970–1979

If the first twenty-two years of the Egyptian-Israeli rivalry were characterized by little or no change in expectations, then the years 1971–1979 began an era of de-escalation and important shifts in this variable and the others that we stress as important. In this chapter, the expectancy model is applied in the context of two periods: the years 1971–1973, which were associated with failed peace initiatives, and the years 1974–1979, which coincided with major diplomatic agreements, such as Sinai I (1974), Sinai II (1975), the Camp David Accords (1978), and the Egyptian-Israeli Peace Treaty (1979). Anwar Sadat emerged as a consolidated policy entrepreneur and confronted his Israeli counterpart, Golda Meir, in the years 1971–1973. Unlike Nasser, Sadat had altered his fundamental attitudes, beliefs, and expectations about the untenable nature of Egypt's rivalry relationship with Israel. On four different occasions from 1971 to 1973, Sadat extended significant peace overtures to Meir, which failed to produce Israeli reciprocity or reinforcement. At the same time, Sadat's efforts occurred without major international third-party pressure. On the Israeli side, Meir was not as formidable a presence in Israeli foreign policy as her predecessor, Ben-Gurion, had been. On the one hand, she maintained conservative beliefs about the feasibility of a diplomatic breakthrough with Sadat, despite his efforts to signal his commitment to a new direction with Israel. On the other hand, she also perceived that her government was constrained by an Israeli public that was unprepared and unwilling to make peace with Egypt even though she was aware that failure to respond to Sadat would in all likelihood result in another war (Shlaim 2001a: 289; Bar-Joseph 2006: 555).

The subsequent discussion also demonstrates that the October War of 1973 was a critical external shock that paved the way for regime changes in Israel. Meir's successors—Yitzhak Rabin and Menachem Begin—were policy entrepreneurs who revised their expectations about Egyptian behavior. Consequently, they generated the necessary diplomatic reciprocity and reinforcement that encouraged Sadat to continue diplomatic negotiations on a bilateral basis with the United States as a key third-party player in the negotiations.

Anwar Sadat and a New Perspective on the Egyptian-Israeli Rivalry

After the June War, two strains of thinking dominated Egypt's national leadership. The first school prevailed during the Nasser years while the second dominated Sadat's rule. The first view held that a political solution was unlikely and that the presence (if not the use) of a strong military force would convince the Israelis to negotiate on the basis of national security interests (Shemesh 2008: 31). The second perspective maintained that a peaceful solution could be achieved especially since the United States would encourage it as a means to reduce the Soviets' growing influence in the region. In addition, the Israelis would be willing to negotiate since negotiation would forestall another military engagement, which would be more likely in view of Egypt's recent arms buildup (Shemesh 2008: 31).

When Sadat moved into power in 1970, he decided to pursue a path that would avoid Nasser's military failures. He believed that Nasser had been overly dependent on the Soviet Union and overly antagonistic to the United States. Moreover, he believed that Nasser was unrealistic to think that Israel could be eliminated through war, because the U.S. commitment to Israel's security was unwavering and the military balance of power favored Israel over the Arabs. Therefore, he believed that the Egyptian-Israeli conflict had to be resolved ultimately through political means (Shemesh 2008: 30).

Consequently, Sadat engaged in a major shift in Egypt's relations with oil-rich, conservative, Arab governments, the United States, and the Soviet Union. He believed that the most productive way to recover territory lost to Israel in the June War was not merely to strengthen Egypt's military via the Soviet Union but to reach out to the United States in the hope that it could pressure the Israelis into negotiating an end to the conflict and the return of Egypt's occupied territories. During Nasser's funeral, Sadat met with the highest U.S.

official in attendance, Elliot Richardson, and expressed his desire to move much closer to the West (Daigle 2004: 3–4). As early as October 1971, Sadat made it clear in a meeting with Secretary of State William Rogers in Cairo that he was serious about removing Soviet troops from Egypt and agreeing to a peace treaty with Israel. From Sadat's point of view, the expulsion of Soviet troops was a signal that he wanted to change U.S.-Egyptian relations and that he hoped for a more even-handed U.S. policy in the region (Daigle 2004: 7–8).

Meanwhile, Sadat sought a rapprochement with Saudi Arabia both politically and economically. Since the Saudis believed that Egypt was weak and poor in 1970, they were less threatened by Egypt's regional policies. They were also encouraged by Sadat's proclamation that he had no political or other ambitions in the Gulf region (Feiler 2003: 130). Nonetheless, the Saudis were threatened by the deepening presence of the Soviet Union in the region and, in particular, Sadat's recent friendship treaty with the Soviets, which was signed on May 27, 1971. Sadat was able to convince the Saudi government that the treaty was no more than a "piece of paper" and that his goal in signing the agreement was to ensure more military aid. Egypt and Saudi Arabia entered into a new understanding that generated increased Saudi financial aid to Egypt beyond the amounts agreed on earlier in Khartoum. Thus the "Cairo-Riyadh" axis was established.

As for the Soviets, Sadat signed that controversial friendship treaty with the Soviet Union in May 1971 largely as a result of Soviet pressure. The Soviets were alarmed by Sadat's decision to purge key pro-Soviet officials in the Egyptian government after Nasser's death and they wanted some public assurance about the strategic relationship between the two governments. Sadat complied largely because he feared that if he refused the Soviets would deprive him of essential aid and arms, weakening his domestic political position and his efforts to rebuild the Egyptian military (Beattie 2000: 86). Nonetheless, Sadat continued to pursue quiet diplomacy with both Saudi Arabia and the United States that signaled his willingness to expel the Soviets in an effort to push the peace process forward.

Sadat's Personal and Political Consolidation of Domestic Power

In addition to Egyptian's deteriorating economy, Sadat faced a major internal problem of establishing his political control over the government. At the time of Nasser's death, "centrists" who favored a close relationship with the Soviet Union and a "statist" economic approach were in control of the highest government offices. These centrists supported Sadat as president because they perceived him to be weak and someone that they could manipulate. However, Sadat was able to neutralize the police and military institutions, eliminate his political centrist opposition, and replace key officials with his own personal allies by May 1971.[1] His allies were mostly rightist in orientation and skeptical of close ties with the Soviets. They were also conservative in their social and religious values, and while they were committed to maintaining the public sector they preferred a more liberal approach in the economic and political systems (Beattie 2000: 37–86).

On the foreign policy front, Sadat confronted enormous domestic pressure to retrieve the Sinai from Israel at a time when Egypt's economy was practically bankrupt. He was aware that his political survival depended heavily on his ability to bring about a change in the status of the Sinai. Because he believed that the more likely path of success was diplomacy, Sadat announced a new diplomatic initiative in search of a peace treaty with Israel from late 1970 throughout 1971.[2] He never found it and by the end of 1972 he had decided on war with Israel.

Sadat's Failed Peace Overtures, 1971–1973

There were four opportunities for a diplomatic breakthrough between Egypt and Israel—all of which were initiated by Sadat to no avail. During this time, the Israeli government pursued two strategies of attrition: one was directed at maintaining the territorial status quo and the other was designed to maintain the political deadlock between Egypt and Israel, denying Egypt any political gains until it accepted Israel's terms for a settlement (Shlaim 2001a: 309).

The first signal to indicate Sadat's interest in a new direction occurred in the context of a UN diplomatic effort led by Dr. Gunnar Jarring in January 1971. Jarring proposed that Egypt enter into a peace agreement with Israel and that Israel withdraw to the former international border between Egypt and Palestine. Sadat replied that Egypt was ready to comply on the basis of UN Resolution 242, which included Israel's withdrawal from the Sinai and the Gaza Strip, a settlement resolving the Palestinian refugee problem, and the establishment of a UN peacekeeping force to maintain the peace. The overture was considered a "breakthrough" since no other Egyptian government had declared publically its willingness to sign a peace treaty. After much discussion within the Israeli cabinet, Prime Minister Golda Meir refused to withdraw to the pre-1967 borders, and much to the disappointment of Jarring and the United States, the Israeli response ended the discussion (Shlaim 2001a: 298–309).

The second signal occurred in February 1971 when Sadat proposed in a speech before his national assembly that the first step toward implementation of UN Resolution 242 could begin with the reopening of the Suez Canal and the partial withdrawal of Israeli forces from its east bank. According to Shlaim (2001a: 301–302), Sadat's initiative suggested that he was shifting the mediation from the United Nations to the United States and from an overall settlement to an interim one. Meanwhile, Egypt approached the U.S. State Department with a follow-up proposal for the feasibility of reaching a settlement to reopen the Suez Canal. Sadat hoped that U.S. pressure on Israel could break the diplomatic deadlock between the two countries.

When Prime Minister Meir responded negatively, Sadat tried again through the U.S. State Department to explain to Israel that he wanted a serious discussion via the United States, not the United Nations. Shortly thereafter in March 1971, the U.S. undersecretary of state, Joseph Sisco, presented a preliminary suggestion that Israel withdraw its forces forty kilometers from the canal, that the evacuated area be demilitarized, and that six months later the canal open to shipping, including Israel's. The resulting agreement would then be a first step toward full implementation of UN Resolution 242 with both sides reviewing the ceasefire after one year. Golda Meir was opposed in principle to the idea of Israeli withdrawal without a peace treaty. She did not like the proposed linkage between the Suez Canal agreement and the full implementation of Resolution 242, and she opposed the presence of Egyptian military soldiers on the east bank of the canal (Shlaim 2001a: 303). Nonetheless, the Israeli cabinet debated the proposal for six weeks and even presented a counterproposal. However, the internal divisions within the Israeli cabinet and the Israeli Defense Forces between dovish officials who supported an interim military withdrawal and hardliners who did not ensured that Golda Meir's position won the day (Shlaim 2001a: 306).

Why was Meir so intransigent? According to Shlaim (2001a: 308), she feared that a partial withdrawal would lead to more pressure for complete withdrawal from the Sinai, and she believed that neither the cabinet nor the

Israeli public would support such concessions to the Egyptians. Shlaim (2001a: 308) also suggests that Meir had a personal issue at stake as well: historically, she did not want to be the first prime minister of Israel to hand over territory to a major Arab rival. Ultimately, she was wedded to the "political, military and territorial status quo."

The next major opportunity for a diplomatic breakthrough occurred in July 1972 when Sadat expelled approximately twenty thousand Soviet military advisors from Egypt without any formal settlement from Israel or promised support from the United States. Although he had several reasons for doing so, two were significant.[3] One reason involved Sadat's preferences for maintaining good relations with the Saudis, who were pressuring him to force a Soviet withdrawal. Sadat was also aware that the Saudis were likely to reward his effort with more much needed financial aid (Feiler 2003: 131–132). At the same time, Sadat hoped that the forced withdrawal would lead to a renewed effort by the United States to pressure the Israelis to negotiate. Sadat viewed the expulsion as a signal of his commitment to moving closer to the West and ending the state of war with Israel (Beattie 2000: 128).

From the Israeli point of view, Sadat's move was perceived as an indication of Egypt's waning military strength and a vindication of Israel's attritional strategy of diplomacy (Shlaim 2001a: 313–314). Meanwhile, the Nixon administration offered little in return except the opening of direct channels of communication to the White House in the context of a meeting between National Security Advisor Henry Kissinger and his Egyptian counterpart, Hafez Isma'il. The meeting would eventually take place on February 25–26, 1973 and would be Sadat's last overture for a diplomatic breakthrough that would avoid war.[4]

The fourth case of missed opportunity occurred at the end of February 1973 during Isma'il's visit with Henry Kissinger. At this time, Isma'il offered what Sadat believed was a far-reaching proposal for a comprehensive settlement of the Egyptian-Israeli dispute. In exchange for Israeli withdrawal from Egyptian territory, Egypt was willing to commit to recognizing Israel's territorial integrity and sovereignty; to end the state of war with Israel; to extend free passage in the international waterways (for example, the Suez Canal); to accept international forces in one or two strategic points in the Sinai; and to end the boycott on third-party goods to Israel. In a second meeting in February, Isma'il also extended Egypt's commitment to prevent acts of terrorism against Israel emanating from Egyptian soil; to accept demilitarization of certain sections of the Sinai; and to proceed with normalization policies once a comprehensive settlement including the Palestinians had been reached. The Egyptians pressed for a speedy bilateral agreement and complete Israeli withdrawal from the Sinai by the end of 1973 (Bar-Joseph 2006: 548).

Kissinger defined the Egyptian proposal as "far-reaching but one-sided" and maintained that it was too close to earlier Egyptian proposals that had already produced diplomatic deadlock. In fact, Kissinger sent a personal message to Sadat saying that Sadat needed to be realistic because he was not in a position to make demands but only more concessions if the United States was to help him. As the defeated party in the negotiations and given Egypt's weak military position vis-à-vis Israel, Egypt was not in a strong position to make demands (Bar-Joseph 2006: 550–551). From Kissinger's perspective, Egypt would not be able to change this strategic reality and with time it would have to make more concessions. Kissinger was wrong. After these meetings, Sadat concluded that a military initiative was the only solution to breaking the deadlock, and by April 1973 Sadat had formed a war cabinet and drawn up war plans with Syria (Beattie 2000: 131; Sela 2000: 47).

Meanwhile, Golda Meir, in her visit with Nixon in March 1973, conveyed her satisfaction with the lack of diplomatic progress and made a concerted effort to secure more U.S. military aid for Israel. She explained that as time passed Israel's control over the occupied territories would eventually be accepted by the international community. She also perceived that Israel's military defense was "impregnable," which meant there was no need for diplomatic change (Shlaim 2001a: 314–315). Essentially, Israel's primary goal in 1972–1973 was to pursue a "diplomatic strategy of attrition": perpetuate the status quo, sustain the political deadlock, and deny the Arabs any political gains until they accepted Israel's terms for a settlement (Shlaim 2001a: 315). Meir and her cabinet stuck to this strategy although they had warnings that Egypt was preparing for war as early as April 1973. Meir and her advisors failed to act diplomatically to avert the coming war (Bar-Joseph 2006: 552–553).

The question remains about how Sadat was able to finance another military confrontation with Israel at a time when the Soviets were backing away from military aid. Sadat's rapprochement with the oil-rich Arab governments played a significant role. In short, Sadat was able to secure additional financial support—without which he would not have been able to launch a war (Feiler 2003: 135). For instance, in mid-December 1972 after Sadat's expulsion of Soviet advisors, Saudi Arabia, Kuwait, and Libya pledged to provide Egypt with a substantial amount of military and financial assistance to be used to purchase arms from the Soviets. From January 1973 to October 1973 when the war began, these Arab governments continued to provide substantial financial support for buying arms, and by October 1973, King Faysal of Saudi Arabia had agreed to an oil boycott in the eventuality of an attack by Egypt against Israel (Feiler 2003: 134).

By the fall of 1973, after six years of war preparation, Sadat had run out of time. At the end of the year, international debt payments had to be met, and Sadat knew that he could not repay them. In addition, substantial domestic unrest began to emerge as students and workers began demonstrating for the end of a military postponement to the Sinai issue (Beattie 2000: 133). Therefore, Sadat changed his military objectives from liberating the Sinai totally from the Israelis as had been planned in 1972 to fighting a limited war for the east bank in the Sinai in 1973. Sadat wanted to achieve a complete strategic surprise early in the war so that his army had time to cross the Suez Canal and grab at least "10 centimeters" of the Sinai. At that point, the superpowers would intervene, establish a ceasefire, and begin a diplomatic process that would end the political stalemate (Sela 2000: 57). Following the conclusion of a new arms deal with the Soviets in late August 1973, Sadat finalized his military plans with President Asad of Syria for a coordinated attack and both agreed to launch the war on October 6, 1973.

The Impact of the October War of 1973: A Shift in Sadat's Strategy and Rabin's Moderation

In the aftermath of the war, Sadat's popularity among the Egyptian public soared as a result of the army's initial success in crossing the canal despite Israel's encirclement of Egypt's Third Army outside Cairo and the intervention of the superpowers to prevent Egypt's military defeat. The "crossing" strengthened Sadat's position domestically, providing him with the space to pursue new domestic and foreign policies. Since he had few economic resources left after the war, Sadat aggressively pursued a peace agreement with the Israelis. Such an agreement would allow him to rechannel money from the military to the economy and regain Egypt's three primary sources of revenue: the Suez Canal, the oil fields of the Sinai, and tourism (Barnett 1992: 129).

In order to achieve such an agreement, Sadat significantly changed his political strategy. First, he sought a negotiated solution with Israel by breaking solidarity with Syria and other Arab confrontation states that maintained a rejectionist stance. Before 1973, Sadat supported a collective Arab effort at securing a comprehensive peace agreement with Israel, but after the war he followed an independent course of action. Moreover, he did so by securing the mediation efforts of the United States, which was better positioned than the Soviet Union to bring about such an outcome. Sadat believed that the Soviet Union was a "second-rate power" with poor technology, while the United States was the "only" superpower that could help him solve Egypt's economic problems and achieve peace with Israel. Therefore, he relied on U.S. mediators, particularly Kissinger, in influencing Israel to negotiate a withdrawal from the Sinai (Beattie 2000: 147–148).

Shifting Egypt's allegiance from the Soviet Union to the United States also served another purpose as well. Sadat hoped to enlist U.S. aid to finance his economic liberalization program at home. The new program, called "al-infitah" (the opening), was launched in 1974 and designed to open Egypt's economy, encourage foreign and domestic capitalist investment, and support private capitalist development—all of which hopefully would rectify Egypt's lack of economic growth, its growing foreign debt, its balance-of-payment crisis, and its depleted foreign exchange reserves (Barnett 1992: 131). Sadat hoped to attract foreign capital and private foreign investment that would lead to new employment opportunities in foreign companies and joint ventures with Egypt's public sector firms. These joint ventures would invigorate Egypt's firms with an infusion of new technologies, capital and management techniques, and a more competitive edge (Beattie 2000: 146).

Sadat knew that if "al-infitah" was to succeed he had to improve relations with those countries whose foreign capital he needed most, in particular, capital from oil-rich countries of the Middle East and North Africa and from the developed countries in the West. If his economic strategy was to work, then Sadat had to convince foreign investors that their investments would be profitable but also safe from destruction in another Middle East war. Therefore, securing peace with Israel became a vital part of Sadat's economic strategy (Beattie 2000: 147). Sadat's strategic shifts in foreign policy, in short, were linked inextricably with his economic reforms at home.

These shifts in foreign and economic policy explain Sadat's willingness to sign two major agreements with the Israelis and terminate Egypt's ties with the Soviets. In January 1974, as a result of U.S. mediation, Sadat signed a disengagement agreement (Sinai I), which required Israel to withdraw from territory on the western and eastern sides of the Suez Canal, which would then became separate buffer zones for Egyptian, Israeli, and UN troops. The agreement also acknowledged that the disengagement was the first step toward a lasting peace in accordance with UN Resolutions 242 and 338. In September 1975, again largely as a result of U.S. mediation, Sadat signed the Sinai II agreement with Israel, which resolved to withdraw from the oil fields on the Suez Canal and the Mitla and Giddi passes in exchange for a U.S. built early-warning station to be operated by U.S. troops.[5]

Lastly, Sadat abrogated the Egyptian-Soviet Treaty of Friendship and Cooperation six months later in March 1976. Sadat's decision to distance himself from the Soviets reflected his concern about Egypt's growing trade dependence on the Soviets for military imports in exchange for Egyptian exports. Such dependence precluded the opportunity of exchanging Egyptian exports for Western technology and inhibited economic growth. Sadat believed that Western technology, Arab petrodollars, and access to international economic institutions such as the International Monetary Fund could deliver the necessary resources to aid Egypt's economic woes. Moreover, not only was the United States positioned to provide access to Western technology and aid, but it was also more likely to bring about peace between Egypt and Israel through its mediation. Hence, Sadat no longer viewed the Soviet Union's economic model or its financial aid to be viable (Barnett 1992: 134).

Sadat's strategy paid off handsomely. In 1975, the United States provided $372 million with approximately $111 million allocated for food aid. After Egypt abrogated the Egyptian-Soviet Friendship Treaty in 1976, U.S. economic assistance escalated to $987 million with $192 million designated as food aid (Beattie 2000: 179). Moreover, the influx of aid from oil-rich Arab states, Japan, and international economic institutions exceeded debt amortization payments by almost $2 billion in 1975 (Beattie 2000: 149). New income also emerged from the opening of the Suez Canal, the Sinai oil fields, tourism, and increased foreign remittances. Sadat's new orientation to the West contributed to new domestic economic wealth and temporarily solved Egypt's liquidity problems.

On the Israeli front, the October War of 1973 weakened the public's support for the Labor-led government. Prime Minister Golda Meir eventually resigned in April 1974 over her handling of national security issues and in particular her government's failure to entertain and detect the possibility that Egypt and Syria would cooperate in a surprise attack against Israel.

Although the Labor Party continued to govern through Yitzhak Rabin's leadership, it was wracked with internal divisions and the loss of its rank-and-file members to conservative and right-wing parties. Rabin and the Labor Party also faced serious economic problems. The costs of the war (that is, one year's GNP) and the increase in oil prices in the wake of the Arab oil embargo produced a decline in economic growth. For instance, Israel's economic growth rates varied between 11.2 percent and 9.7 percent in the 1950s and 1960s. Between 1973 and 1980, however, those rates decreased to 3.2 percent. Meanwhile, the defense burden surged dramatically after the war to as much as 28 percent of the GDP (or 16 percent subtracting U.S. aid and other external aid) (Y. Levy 1997: 146). Concurrently, in 1974–1975 global inflation associated with the oil embargo increased Israel's balance-of-payments deficits to $3–4 billion from their typical level of $500,000 to $1 billion before 1973, and Israel's national debt escalated from $980 million in 1965 to more than $9 billion in 1978 (Crittenden 1979: 1007).

As Israel became more dependent on the United States to cover its growing defense burden (as much as 40 percent of its defense budget) and balance-of-payments deficits, the Labor government increased national taxes, cut subsidies, and devalued the Israeli pound (Crittenden 1979: 1007; Y. Levy 1997: 146). These austerity measures eroded Rabin's public support among the lower classes, who were becoming attracted to more conservative and religious parties (Barnett 1992: 189).

In addition to Israel's economic woes, Rabin's government was weak, vulnerable to internal disputes, and cautious about its foreign policy. Rabin, himself, advocated an ambiguous policy based on "continuity and change" and he was evasive about what territorial compromises he would be willing to make to secure peace with Egypt. He preferred moving slowly, playing for more time when Israel would be economically and militarily stronger. He refused to allow Egypt and other Arabs to believe that Israel would make concessions from a position of weakness (Shlaim 2001a: 326–327). So, Rabin pursued a "wait and see" approach, demanding that a second disengagement agreement be premised on the principle of reciprocity; that is, that Egypt end its state of belligerence as a precondition for Israel's withdrawal from parts of the Sinai. Sadat, of course, refused to make any such formal announcement, and even Kissinger, who was mediating the terms of what would eventually be the principles of the Sinai II agreement, argued that Israel's position was unreasonable (Ben-Zvi 1993: 94).

At this point, the United States played a critical role in influencing Rabin to accept a more modest plan restricted to a unilateral military withdrawal from parts of the Sinai. In compensation for this concession, President Ford and Secretary of State Kissinger signed a "memo of understanding" that committed the United States to be responsive to Israel's military, economic, and energy demands. In other words, the United States agreed to provide large-scale economic and military aid (approximately $1.5 billion in military credits, plus about half that amount in economic aid for the following year), and advanced weapons. In addition, the United States made commitments to protect Israel from other major power threats, supply oil if it should be denied access, refurbish it with more advanced weapons over time, maintain the U.S. policy of nonrecognition toward the PLO, and consult closely with Israel on any future Middle East proposals (Ben-Zvi 1993: 101–102). The cost of the agreement to the United States was approximately $4 billion annually for the next three years, or 200 percent above the existing level of U.S. aid to Israel (Shlaim 2001a: 338). In return for relinquishing one-seventh of the Egyptian territory along the Suez, including the strategic passes and oil fields, Rabin gained breathing space to rebuild Israel's military capabilities, deepen the U.S.-Israeli relationship and negotiate a deal that was independent of a comprehensive Middle East peace agreement involving Syria and Jordan.

In sum, the October War exacted enormous costs on both the Egyptian and Israeli economies. Six years of war preparation left Sadat with a debt-ridden economy that required a rapid infusion of aid and investment. Discarding his alliance with the Soviets and striking out on his own, Sadat sought U.S. diplomatic and economic aid in securing two major disengagement agreements with the Israelis. The October War, in turn, increased Israel's economic vulnerability and its dependence on the United States for external aid. Consequently, after much pressure from the United States, Rabin participated in the Sinai agreements, but only after he had obtained strong commitments from the United States to provide aid, arms, and diplomatic support in the future. Although Sadat's reorientation of his foreign policy played a key role achieving the Sinai I and Sinai II agreements, Sadat's success is due largely to the active mediation role of the United States and its economic and military largess to both Egypt and Israel.

Economic and Political Pressures Prior to 1978

Egypt and the Food Riots of January 1977

Although economic liberalization improved Egypt's economy, foreign debt levels accelerated from 1973 to 1976 and government subsidies increased more than thirteen-fold (Beattie 2000: 207). In fact, government borrowing in 1975 had tripled that of 1973 and constituted almost one-third of the GDP (Karawan 2005: 329). By 1977, the foreign debt was $12.2 billion; $2.1 billion in short-term loans with high interest rates, $2.5 billion in Arab financial deposits, $3.4 billion in long-term loans, and $4.2 billion from Eastern Europe. Debt servicing made up 50 percent of Egypt's export revenues, and including the military debt it amounted to 70 percent (Karawan 2005: 329).

The debt levels made it increasingly difficult to get new lines of credit especially from the Arab oil-producing countries, which began to dissociate themselves publically from Sadat's government, especially after the Sinai II agreement.[6] Refusing to commit to an "Arab Marshall Plan," the Gulf states tied any future financial aid to Egypt's commitment to undertake austerity reforms recommended by the International Monetary Fund (Feiler 2003: 138). One of these austerity measures involved the cutting of government subsidies, which constituted 33.6 percent of government spending in 1976 (Waterbury 1983: 214). In light of the pressures from the International Monetary Fund and Egypt's creditors, Egyptian officials reduced subsidies on thirty commodities, which in turn generated increased prices for basic foodstuffs. (Beattie 2000: 208; Rivlin and Even 2004: 42). The increase would have cut the purchasing power of about 30 percent of the population by 25 percent or more and triggered an avalanche of other price increases that would have raised the costs of living beyond the means of urban dwellers (Karawan 2005: 330). Immediately, the government's decision generated two days of violent protests and demonstrations by industrial workers, government employees, pension recipients, and the urban poor. At least fifty to seventy-nine people were killed with another six hundred to eight hundred wounded by the time the government rescinded its austerity measures three days later (Beattie 2000: 208; Rivlin and Even 2004: 42).

After the riots, government expenditures on subsidies continued to rise. By 1979, these expenditures had increased to 1.27 billion Egyptian pounds (about $1.85 billion), constituting almost 20 percent of the national budget in that year (Rivlin and Even 2004: 42).[7] Meanwhile, government spending on the military stayed at the same levels from 1973 to 1977, and in 1976 constituted as much as 37 percent of the GNP. According to one estimate, Egypt spent three times the amount on arms from 1975 to 1981 as it had in the previous twenty years (Barnett 1992: 130).[8] Consequently, Egypt was trying to meet social welfare, development, and defense needs at a time when its revenue base was not expanding rapidly enough.

The riots and demonstrations in January 1977 revealed the depth of unrest in urban districts all over Egypt. Sadat was aware that the government's policies were generating deep grievances among Egypt's urban poor and stimulating the mobilization of Islamist, Nasserite, and leftist opposition groups. After the riots, Sadat designated 1977 as the "year of settlement" with the Israelis. The increasing political stalemate between Egypt and Israel since the Sinai II agreement was unbearable for Sadat, who desperately wanted a final agreement that would restore Egypt's access to revenues from the Suez Canal and tourism as well as link the country to Western foreign aid and investment. Rather than asking the Egyptian public to suffer from economic belt-tightening policies, Sadat perceived that the easier, safer alternative was to look for international sources of financing which could be obtained only with a final peace agreement with Israel (Beattie 2000; Barnett 1992).

Thus, Sadat embarked on his famous trip to Jerusalem in November 1977 at time when he believed that there was no alternative to achieving peace with Israel. By speaking in front of the Knesset, Sadat conveyed his implicit recognition of Israel's sovereign legitimacy in the Middle East and signaled a significant change in Egypt's foreign policy. Sadat made this overture because he desperately needed a peaceful resolution so that Egypt could receive the material, factories, and food, quickly and cheaply on predictable terms, in order to revitalize his economy. He believed that the United States and the West could meet these needs and the loss of Arab support in the context of a major peace agreement with Israel was less significant (Stein 1993: 86).

Israel, the National Election in 1977, and Begin

The Likud Party's victory in the parliamentary election in May 1977 ejected the Labor Party from office for the first time in Israel's history. Leading the Likud Party, Menachem Begin emerged as prime minister after piecing together a narrow right-wing coalition of conservative and religious parties. Begin's foreign policy priorities differed significantly from those of the previous Labor governments, particularly on the issue of the occupied territories. Although he was keen on continuing Labor's search for a peace agreement in the Middle East, Begin was unwilling to negotiate any deal that relinquished Israel's hold on the occupied territories. Begin and the Likud Party were ideologically committed to the notion that the West Bank and the Gaza Strip belonged under Jewish control. Shortly after forming his new government, Begin pledged to build Jewish settlements in these areas, although he had no immediate plan of annexing them.

Begin's foreign policy strategy rested on achieving a peace agreement with Egypt in order to remove it as a rival and ensure peace on one of its borders. However, he was unwilling to exchange any territory outside of the Sinai Peninsula. What Begin sought was not so much an exchange of Sinai territory for nonbelligerency but an exchange based on Israel's continued control over the West Bank through a proposed Palestinian autonomy plan that would be endorsed by both Egypt and the United States (Sachar 1981: 258; Shlaim 2001a: 362).

Shortly after his election, Begin pursued the option of a bilateral agreement with Egypt by secretly signaling Sadat that Israel was interested in direct peace talks. Choosing to ignore the multilateral talks cosponsored by the United States and the Soviet Union at the Geneva Peace Conference, Sadat announced on November 9, 1977, his willingness to travel directly to Israel to negotiate a peace deal that would return the Sinai to Egyptian sovereignty. In addition, he reiterated his position that he wanted a comprehensive peace that would lead to Israel's full withdrawal to the borders of 1967 and Israel's recognition of the Palestinians' right to sovereignty. Four days later, Begin invited Sadat to Jerusalem to conduct talks that would hopefully culminate in a permanent peace between both countries.

During Sadat's visit to Jerusalem, Sadat and Begin agreed to three principles: the rejection of war between the two countries; the formal restoration of Egyptian sovereignty over the Sinai Peninsula; and the demilitarization of most of the Sinai, with a limited Egyptian military presence in the area adjoining the Suez Canal, including the Mitla and Gidi passes (Sachar 1981: 267; Shlaim 2001a: 361). Sadat's demands for total Israeli withdrawal and Palestinian recognition were unresolved.

Begin's strategy can be understood in the context of three internal pressures—two of which emanated from domestic constituencies. Gush Emunim ("Bloc of the Faithful") emerged soon after the October War in 1974 as a political movement of young, fundamentalist Israelis who demanded that Israeli law and Jewish settlements be established in the West Bank. Its members staged protest demonstrations that included illegal Jewish settlements in the occupied territories, violent attacks on local Palestinians, and opposition to any significant peace negotiations with Egypt. Despite their radical tactics, Gush Emunim's interests were represented in all the secular and religious parties of the right in Israeli politics—the very groups with whom Rabin formed a government after the national election in 1977. In fact, Begin and the Likud Party were especially supportive of and sympathetic to Gush Emunim's demands that Judea and Samaria remain in Israel's permanent control regardless of Palestinian claims, which explains Begin's unwillingness to negotiate with Sadat on the future of the occupied territories (Y. Levy 1997: 156).

In addition to Gush Emunim, the Peace Now movement, composed initially of retired military officers, appeared in 1978 and soon expanded to include academics, professionals, and semi-professionals. Peace Now activists mobilized in order to put greater pressure on Begin to advance a peace agreement with Egypt that involved territorial compromises. In addition, Peace Now opposed the expansion of Jewish settlements in the occupied territories. Throughout the spring and summer of 1978, activists organized massive demonstrations, met with government and party officials, and arranged media campaigns to maintain the pressure on Begin and his government. On September 12, 1978, the day before Begin left for Camp David and direct talks with Egypt, Peace Now activists, numbering over one hundred thousand citizens, marched on Tel Aviv's town hall, pushing Begin to make the necessary concessions to achieve a peace treaty (Bar-On 1996: 112).

Another important factor influencing Begin's strategy was Israel's deteriorating economy, especially the problems associated with the increasing balance-of-payments deficit, growing hyperinflation, and rising economic dependence on U.S. aid. Begin pursued an economic liberalization policy, which loosened controls over foreign currency, allowing the Israeli pound to find its own level in the foreign exchange market. In addition, Begin slashed export subsidies and reduced subsidies on basic commodities. His reforms were intended to attract foreign investment, encourage capital flows back into Israel, force Israeli industry to be more competitive without domestic protection, and stimulate the sale of Israeli goods abroad through the devaluation of the Israeli pound. Begin surmised that if the plan worked, Israel's economy would not need the $1 billion of U.S. aid that it had been receiving since 1975 (Crittenden 1979: 1007–1009).

Largely due to Begin's unwillingness to cut defense-related expenditures or social and welfare services, these reforms, along with the rising national debt, brought about spiraling inflation. By 1985, inflation stood at an annual rate of 400 percent (Y. Levy 1997: 147). Consequently, Israel's dependence on the U.S. aid had increased dramatically. For instance, in the five years after the October War of 1973, the United States transferred more than $10 billion in official assistance (Crittenden 1979: 1013). This aid made it possible for Israel to avoid raising taxes and diverting civilian expenditures. In addition, it helped Israel deal with its national debt, pay off its war losses, and cope with the increasing costs of energy following the Arab oil boycott after the October War. In short, Israel's external dependence meant that U.S. aid was critical to maintaining Israel's standard of living, beyond the military weapons that the United States provided (Y. Levy 1997: 148). Increasing dependence on the United States, however, meant that Begin was highly sensitive to maintaining good relations with President Jimmy Carter, who was pressuring him to make significant concessions on both the Sinai and the Palestinian territories.

U.S. Mediation and the Camp David Accords of 1978

In the year preceding the Camp David meetings in September 1978, the negotiations between Egypt and Israel reached a stalemate. The knottiest issue centered on Sadat's demands for full restoration of the Sinai and national self-determination for the Palestinians. Begin countered with a Palestinian autonomy plan that would involve self-government for the West Bank and the Gaza Strip while Israel maintained internal security. After a five-year interim period, Israel would review the situation. Angry about the recent construction of Israeli settlements in Rafah (Sinai), the Egyptians demanded total Israeli withdrawal from the Sinai, full Palestinian sovereignty, and no separate peace agreement (Shlaim 2001a: 364–367).

In August 1978, President Carter invited Begin and Sadat to attend a summit meeting at Camp David. Both leaders accepted the invitation without any preconditions. When Carter, Begin, and Sadat arrived at Camp David, the focus centered on a bilateral peace agreement dealing with the normalization of Egyptian-Israeli relations, the removal of Israeli settlements in the Sinai, the scope of the demilitarization of the Sinai, and an agreement on a general principle on the right to Palestinian self-determination and a linkage between normalization and the future of the Palestinians (Stein 1993: 87–88).

Stein (1993: 89) argues that at this point Begin and Sadat were aware of the major political costs they would incur if they emerged from Camp David without an agreement. Sadat needed Israeli withdrawal from the Sinai in order to be able to address the economic problems in Egypt. The economic costs were not as dire for Begin, but he was conscious of the necessity of maintaining good relations with the United States. Begin knew that maintaining good economic and political support from the United States was critical for Israel because of its isolation in the international community and its dependence on U.S. aid (Stein 1993: 89). Similarly, Sadat was not immune to Carter's warnings that if Sadat left Camp David without securing an agreement he could lose U.S. support as well, in particular, its aid, investment, technology, and continued diplomatic pressure on Israel. Therefore, Carter was able to move both Begin and Sadat to accept two frameworks for peace after thirteen days of intense negotiations.

The negotiations at Camp David ended at the White House with the signing of the Camp David Accords on September 17, 1978. The success at Camp David was largely due to Carter's dogged determination to see the negotiations conclude successfully. Without his involvement and his efforts to reconcile Begin's and Sadat's negotiating positions, the Camp David Accords would not have been signed. Begin and Sadat were also aware that walking away from the negotiations would assure them of being blamed for the failure, of disappointing a U.S. president, and consequently, of ensuring a less than friendly relationship with United States in the future (Stein 1999: 252).

The Camp David Accords were not so much a treaty as a promise of two negotiating tracks: one dealing with the future of the Palestinians in the

West Bank and the Gaza Strip and the other relating to the Egyptian-Israeli agreements on the Sinai. In the coming year, the framework for Palestinian sovereignty or autonomy would not come to fruition. However, the second framework laid the groundwork for the Egyptian-Israeli treaty that would be signed by Begin and Sadat six months later in March 1979. At Camp David in 1978, the Israelis agreed to a full withdrawal from the Sinai, including the evacuation of Israeli citizens in the area within three years and the return of the oil fields in the western Sinai. In return, Egypt would restore normal diplomatic relations with Israel, guarantee free passage through the Suez Canal, and restrict the number of its forces along areas close to Israeli borders. Both Sadat and Begin were able to consolidate their governments' ratification of the treaty without much opposition due to their relative autonomy within their governments. While Sadat had a strong autocratic control over his political environment, Begin was the undisputable leader of the Likud Party and was able to rally domestic support for his diplomacy (Stein 1993: 93).

The agreements of 1978 and 1979 also resulted in the United States committing several billion dollars worth of annual subsidies to both Egypt and Israel, much of which continue to this day. The United States provided a total of $7.5 billion to both Egypt and Israel in 1979 (Sharp 2009: 19). Egypt also received assistance from the World Bank, West Germany, and Japan in exchange for the political agreement. Their aid totaled $1 billion in 1979, and in the same year Western businessmen invested about $500 million in the Egyptian economy (Feiler 2003: 212).

The Application of the Expectancy Revision Model of Rivalry De-escalation

Table 4.1 shows the critical variation in the model's variables. In the first period of Sadat's peace initiatives, the absence of all but two of the six independent variables accounts for the lack of rivalry de-escalation. In contrast, the post-1973 era is associated with the presence of five of the six variables of our model across the administrations of Sadat, Meir, Rabin, and Begin and explains the de-escalation of the rivalry.

War shocks played a considerable role in changing the attitudes and expectations for Anwar Sadat, Yitzhak Rabin, and Menachem Begin, but not for Golda Meir. Sadat's expectations for peace changed after the June War in

Table 4.1. Modeling Outcomes for the Egyptian-Israeli Rivalry, 1970–1979

Variable	*Failed peace initiatives, 1971–1973 Sadat (Egypt)*	*Meir (Israel)*	*Successful peace initiatives, 1974–1979 Sadat (Egypt)*	*Rabin (Israel)*	*Begin (Israel)*
Shocks	Present	Present	Present	Present	Present
Expectancy revision	Present	Absent	Present	Present	Present
Consolidated and new policy entrepreneur	Present	Absent	Present	Present but not consolidated	Present
Third-party pressure	Absent	Absent	Present	Present	Present
Reciprocity	Present	Absent	Present	Present	Present
Reinforcement	Not applicable	Absent	Present	Present	Present
Outcome	No change	No change	Some de-escalation	Some de-escalation	Some de-escalation

1967, but it would take another war with Israel in 1973 before the expectations of Rabin and Begin would also change.

Sadat's expectational shifts are discernible in the aftermath of the war in 1967. The war diminished Egypt's regional role, but more significantly, it pushed Egypt to the edge of financial bankruptcy. As a consequence, Sadat sought to link the restoration of Egypt's deteriorating economy with the termination of hostilities with Israel. Sadat believed that securing a peace treaty with Israel would lead to a recovery of lost government revenues from the Suez Canal and also open the door to new sources of economic aid, technology, and investment from the United States and Western Europe.

While the June War had less effect on changing Israeli expectations about peace with Egypt, the October War had considerable impacts on both Rabin and Begin. Rabin signed the Sinai I and Sinai II disengagement agreements, the latter of which laid the foundation for the Egyptian-Israeli Peace Treaty of 1979. Nonetheless, Rabin's attitudes and expectations about Egypt had not changed so dramatically that he viewed these agreements as irrevocable steps to a final peace agreement with Egypt. Rabin's moderation from 1974 to 1977 stemmed from his desire to consolidate Egypt's new pro-U.S. orientation, to place a wedge between Egypt and Syria in order to eliminate a future threat of a two-pronged war against Israel, and to maintain good relations with the United States, which pressured him to commit to these agreements (Inbar 1999: 21).

Rabin's successor, Menachem Begin, also saw the merits of negotiating a peace settlement with Egypt in the aftermath of the October War for the purpose of removing a significant rival on one of Israel's borders. However, he was willing to do so only under terms that would ensure Israel's continued control over the West Bank and the Gaza Strip. Begin's perceptions were shaped by his recognition of the necessity of maintaining good relations with the United States at a time when Israel was becoming more and more economically and militarily dependent on it. Faced with a deteriorating economy, international pressure from the United States, a mobilized Peace Now Movement that advocated for a treaty with Egypt, and a highly mobilized settler movement that demanded permanent access to the West Bank, Begin negotiated a peace treaty with Egypt in 1979 largely on Israel's terms.

From 1971 to 1979, Sadat was a policy entrepreneur who was able to advance his new foreign policy goals by consolidating his political control inside and outside of his government. On the Israeli side, however, Golda Meir was neither a policy innovator nor one who had established complete control over her cabinet. Rabin was an innovator in that he was willing to pursue peace negotiations with Egypt, but he was constrained by a divided cabinet and domestic public opinion that was shifting in a more conservative direction. Hence, Rabin was cautious and unwilling to take dramatic steps toward peace with Egypt. Begin, in contrast, exhibited both traits (Stein 1993: 91). He was willing to move in a more controversial direction by signing the Camp David Accords in 1978 because of his personal and political control over the Likud Party, which dominated his cabinet, and his ability to mobilize support in the Knesset. Begin successfully stitched together a coalition of support from within the Likud and the Labor parties to gain a majority of the votes in the Knesset for ratification of the accords. Begin also enjoyed widespread public support for the accords as well (Shlaim 2001a: 377). In 1979, Begin achieved the ratification of the Egyptian-Israeli Peace Treaty with even more support from the Knesset.

With the exception of Golda Meir in the years 1971–1973, there were many cases of reciprocity and reinforcement between Sadat and the Israeli prime ministers. During the years 1971–1973, Sadat initiated a series of overtures on four different occasions—all of which were rebuffed by Meir. However, after the October War, renewed efforts by Sadat to convey his serious intentions to end Egyptian hostility were received in a more positive manner by both Rabin and Begin. The step-by-step or graduated peace negotiations between Egypt and Israel, which were mediated by Kissinger between 1973 and 1975, created room for reciprocity and reinforcement to lay the groundwork for the agreements in 1978 and 1979. Probably, the most dramatic and obvious effort to signal such intentions was Sadat's visit to Jerusalem and his talk before the Knesset in 1977. This event was well received within Israel and paved the way for Begin's commitment to enter negotiations with the United States and Egypt at Camp David.

Mediation by Secretary of State Henry Kissinger during the years 1973–1975 and President Jimmy Carter in the years 1978–1979 were critical to the de-escalation process. At the end of the October War in 1973, Kissinger negotiated a series of partial agreements that first resulted in a disengagement of Egyptian and Israeli military forces in 1974; in 1975 Kissinger negotiated a broader agreement that led to the withdrawal of Israeli forces from strategic areas in the Sinai in exchange for a nonbelligerency pact from Egypt. This agreement was the first time that Israel had agreed to a significant withdrawal of forces since the June War of 1967 and the first time that an Arab leader supported a nonbelligerency agreement (Stein 1985: 332).

In 1978–1979, Carter pursed a similar mediating style. He negotiated bilaterally with Israel and Egypt directly and then used those discussions to forge an agreement between Israel and Egypt.

The United States was successful in negotiating these agreements for several reasons. First, the United States acted as a guarantor of the observance of the treaty to both sides, thereby reducing the risks of concessions (Touval 1982: 225–320). For instance, during Kissinger's negotiations, the United States offered to intervene and consult as to appropriate action should one of the parties violate their agreements. The United States also promised Israel that it would veto any UN Security Council resolution that would adversely affect the agreement. Kissinger also committed the United States to providing aerial reconnaissance of the demilitarized zones in the Sinai and transmitting the data to both sides (Stein 1985: 343).

Similarly, Carter also promised to guarantee that both parties would observe the treaty. Carter went so far as to promise that the United States would consult with Congress in the event that there was a violation of the treaty and that it would take appropriate actions to ensure compliance with the treaty. Moreover, the United States promised to support appropriate actions by Israel should the treaty be violated and to oppose any UN resolution that could adversely affect the treaty. Like Kissinger, Carter also committed the United States to monitoring the implementation of the limitation of force arrangements and to transmit the information to both sides (Stein 1985: 343).

Another reason that U.S. mediation helped to bring about the deescalation of the Egyptian-Israeli rivalry is the role that the United States was willing to play as a "benefactor extraordinaire" by committing significant amounts of long-term economic and military aid to Egypt and Israel. Hence, Kissinger and Carter were willing to "guarantee, insure and reward" both countries if their leaders were willing to make concessions. The United States also acted coercively by threatening the loss of good future relations should one or both parties be unwilling to make important concessions that would produce an agreement. Hence, the United States had enormous bargaining leverage over the participants. Of course, U.S. mediation was successful because it was timed with the rise of Egyptian and Israel leaders who were pessimistic about the benefits that were or could be derived from past or future conflict. In sum, Sadat, Rabin, and Begin considered military force to be an unappealing option because of the increasing domestic political and economic strains associated with each war. As Stein (1985: 345) argues,

"Their estimates of likely economic and military loss, of adverse political consequences at home, and continued military tension were critical" in redefining the parameters of the Egyptian-Israeli rivalry. Despite the skill, motivation, persistence, coercive pressure, and wealth of the United States, Kissinger and Carter would not have been able to accelerate the de-escalation process without a significant prior reorientation of the strategic thinking of the key actors.

The expectancy model of rivalry de-escalation brings together six critical variables that worked together to bring about the rethinking of political leaders involved in an intense rivalry. In the years 1952–1970, with the exception of internal and external shocks, none of the remaining variables lined up on both sides of the rivalry. So, the presence of shocks alone is neither a necessary condition nor a sufficient one for the de-escalation of a rivalry. The years 1971–1979, in contrast, can be divided into two periods: the years 1971–1973, in which Egypt sought a rivalry de-escalation that was unsuccessful, and the years 1974–1979, in which Egypt's initiatives finally paid off in bringing about a de-escalation. The model again applies quite well. It shows that despite the presence of shocks in the first phase (see Table 4.2) none of the remaining variables were present on both sides of the rivalry in the years 1971–1973. However, in the subsequent period, all the critical variables (with the exception of Rabin's consolidated leadership) were present on both the Egyptian and Israeli sides. These cases remind us that

Table 4.2. Shocks in the Egyptian-Israeli Rivalry, 1970–1979

Time frame of shocks	*Shift in external threat*	*Shift in competitive ability*	*Shift in regime orientations/ strategies*	*Economic resource crisis*
1970–1973		Despite Egyptian defeat, the 1973 war enhanced Egyptian self-esteem and Israeli perception of strategic vulnerability.		Economic problems in both Egypt and Israel led to both economies increasing their dependence on U.S. aid.
1974–1979				

shocks do not always have an immediate dampening effect on rivalry hostilities and in those cases where shocks do indeed bring about a new strategic orientation, such reorientations do not occur simultaneously on both sides of the rivalry. Lags in reaction can occur. The same can be said for the other key variables of the model. Hence, rivalry de-escalation would appear to be the consequence of good timing by a range of actors both inside and outside the rivalry. Policy changes are most likely to be achieved when the "stars"—in this case, the presence of the critical variables—are in alignment. When they are not, considerable diplomatic frustration is likely to be the norm.

CHAPTER 5

The Israeli-Syrian Rivalry, 1948–2000, and the Israeli-Palestinian Rivalry, 1980s and Early 1990s

While the Egyptian-Israeli rivalry de-escalated to some less than full extent, sometimes described as a "cold peace," Israel's relationships with Syria and Palestinians, although also subject to windows of opportunity for de-escalation at various points in time, have failed to make much of a dent in either of the other key antagonistic relationships that have been central to Arab-Israeli conflict. In this respect, the comparison between these two cases and the Egyptian-Israeli outcome offers an excellent opportunity to see whether or to what extent our expectancy model can distinguish among the cases. This chapter first examines the ups and downs of the Israeli-Syrian relationship through 2000[1] and then looks at the one time that the Israeli-Palestinian relationship almost changed in the early 1990s.

The Israeli-Syrian Rivalry

Shocks figure prominently in our understanding of the de-escalation process involving Israel and Syria between 1992 and 2000. In the Syrian case, acceptance of the loss of competitive advantage is particularly important. Repeated defeats in warfare with Israel between 1948 and 1982 and the loss of the Soviet Union's patronage at the end of the 1980s resulted in altering Syria's traditional strategies vis-à-vis Israel. On the Israeli side, declining threat perceptions were more important. Changes in internal leadership that produced policy entrepreneurs (starting in 1992), and the emergence of new threats from Iran and radical Islamic political groups that used terrorism produced intermittent shifts in Israel's long-term strategies toward Syria.

In the analysis that follows, the impacts of these shocks on the decision-making processes in Israel and Syria are investigated. The de-escalation process unfolds with each successive shock from 1948 to 2000. In some instances, warfare and changes in leadership solidified hard-line policies, but in others they had significant influence on bringing forth new attitudes and diplomatic flexibility. How and why this happened is explained below. Meanwhile, the de-escalation process from 1992 to 2000 did not result in a rivalry termination outcome. Why that did not happen will be discussed as well. A preview of the answer centers primarily on the absence of reciprocity and reinforcement in the de-escalation process, as well as domestic political constraints and contrasting views about what the peace process would ultimately yield in the long run.

The Arab-Israeli War of 1948 and a Syrian Peace Overture in 1949

After Prime Minister David Ben-Gurion proclaimed an independent Israel on May 14, 1948, Syria along with Egypt, Jordan, Iraq, and Lebanon initiated a war against Israel on May 15. After defeating Arab forces, Israel signed armistice agreements with all the parties by the end of 1949. The most contentious issue between Israel and Syria during this time involved a territorial dispute over the Jordan River and Lake Tiberias.

During the war of 1948, Syria moved down from the Golan Heights to consolidate its control over the northeast shoreline of Lake Tiberias or the Sea of Galilee, the east bank of the Jordan River (north of Lake Tiberias), and a small slice of Israeli territory west of the river. Israelis believed that Syria had violated the UN partition plan of 1947 as well as the international border arrangement of 1923 between Syria and Palestine, which placed Lake Tiberias and the Jordan River in Israeli territory.[2] The dispute was resolved in the armistice agreement of July 1949, which produced three demilitarized zones in the area under the supervision of a UN-Israeli-Syrian commission. The issue of political sovereignty of the area would be resolved in a future peace settlement (Neff 1994; Rabinovich 1991).

Prior to the armistice, Colonel Husni Za'im moved into Syrian leadership in the aftermath of a coup in March 1949 that was supported by the

United States. Preferring a political solution rather than a military one, Za'im extended a generous peace offer to Prime Minister Ben-Gurion during the spring of 1949. Za'im was prepared to sign a final peace agreement that would include open borders between the two countries, immediate diplomatic exchange of ambassadors, military and economic cooperation, and resettlement of three hundred thousand Palestinian refugees in Syria, while Israel would make territorial concessions along Lake Tiberias. Za'im also expressed his willingness to meet with Ben-Gurion to discuss the details. According to Shlaim (2001a), Ben-Gurion ultimately turned down the deal since Syria was unwilling to give up territory that it had occupied during the war of 1948.

Why did Za'im extend such an offer? As a pragmatic politician, Za'im had little commitment to pan-Arabism and was more concerned with consolidating his control in Syria and directing Syria into an independent regional player in the Middle East. Since he perceived the conflict with Israel as a regional issue—rather than an ideological one—he sought a deal with Israel in order to neutralize a future military threat, gain strategic advantages along the Syrian-Israeli border, and use his relationship with Israel as a possible counterbalance to Iraq. In addition, Ma'oz (1995: 24) notes that Za'im hoped to secure a stronger relationship with the United States for financial assistance through a political settlement with Israel.

Israel's Ben-Gurion rejected the offer immediately. He demanded that the Syrians withdraw their forces from Israeli territory before a final peace treaty could be negotiated. Za'im then extended new terms through the United Nations that again included Syria's absorption of three hundred thousand Palestinian refugees and a peace settlement that was based on the old international borders, a significant change from his earlier position. Za'im lobbied intensely for private meetings with Ben-Gurion to resolve their differences. Again, Ben-Gurion rejected Za'im's offer.

Despite Ben-Gurion's near monopoly over foreign policymaking at the time, his position was endorsed by other Israeli elites and the army. Ben-Gurion's inflexibility was due to his preoccupation with large-scale domestic problems of state building and his unwillingness to cede territory to hostile Arab neighbors in exchange for formal peace treaties. Ben-Gurion was not in a rush to sign a peace treaty because he believed that Israel's bargaining position vis-à-vis the Arabs would only improve with over time. Meanwhile, he believed (Shlaim 2001a: 51) that the price of yielding territory to the Arabs while simultaneously accepting a return of large numbers of Palestinians was too high relative to the benefits that could be derived from formal peace treaties.

1950–1951: A Military Coup in Syria; Another Peace Initiative

The next major overture to the Israelis came from Adib Shishakli, who had secured his leadership in Syria by 1951 as a result of a coup in December 1949. In May 1952, Shishakli extended an offer not unlike Za'im's proposal: in exchange for a peace treaty with Israel, Syria would absorb half a million Palestinian refugees. In addition, Israel would divide the demilitarized zones, which would allow Syrian access to both the Jordan River and Lake Tiberias. Ben-Gurion refused once again to make any territorial or water rights concessions (see Slater 2002: 87).

Shishakli's offer occurred in the context of a series of border clashes between Israeli and Syrian forces. In early 1951, the Israelis were determined to secure their control over the demilitarized zones by establishing "facts on the ground." In that effort, they removed Arab residents from these areas, replacing them with paramilitary agricultural settlements. As settlers began cultivating the land, Syrian troops fired on them, provoking retaliatory raids from the Israelis. By 1952 largely as a result of these clashes, the demilitarized zones were effectively partitioned between Israel and Syria. Israel controlled the west side of the Jordan River, while Syria maintained control over the northeast corner of Lake Tiberias and the east bank of the Jordan River. As noted in Slater (2002: 88) this border remained in effect until the Six Day War of 1967, when Israeli forces overran it.

Shishakli's peace offer to Israel was motivated by his desire to subordinate the problem of the Israeli conflict while focusing his efforts on socioeconomic reforms and eliminating sources of internal instability through his efforts to modernize and build the Syrian army. Shishakli pursued these aims by establishing better relations with the United States in the hopes of receiving both military and financial aid. In return for U.S. help, Shishakli was willing to make peace with Israel by settling for a nonbelligerency agreement that did not entail political and economic relations but did ensure the end of border clashes and the absorption of large numbers of Palestinian refugees.[3]

Although the Israelis ultimately rejected Shishakli's offer, he was never really in a position to deliver such a deal because large segments of both the public and the political elite harbored deep animosity and distrust of the Israelis. Syrians were threatened by Israel's growing military power and its efforts to expand its control over land adjacent to its borders. Although Syrian leaders frequently engaged in external threat outbidding in their efforts to sustain their political positions and ideological views, they were still fearful enough to suggest defense pacts with other Arab countries such as Egypt, Lebanon, Jordan, and Saudi Arabia (Ma'oz 1995: 34).

The Israelis, meanwhile, equally distrusted the Syrians. In fact, Israelis regarded the Syrians as more threatening than their other Arab rivals. From the Israeli perspective, the Syrians were highly nationalistic and virulently anti-Zionist. The Syrians repeatedly invoked the necessity of destroying the Jewish state and its citizens in order to protect Syria's territorial integrity and pursue its pan-Arab ambitions. Finally, Ben-Gurion was intransigent on making territorial concessions to the Arabs as well as sharing the waters of the Jordan River and Lake Tiberias (Ma'oz 1995: 34–35; Slater 2002: 87).

Despite Za'im's and Shishakli's pragmatic offers to the Israelis, the probability of any serious peace agreement emerging from these early discussions was negligible in light of the zero sum attitudes held between the two governments. Ma'oz (1995: 36) suggests that the war of 1948 was still too recent in the collective memories of the two governments. Mutual fear, distrust, and prejudice overshadowed the possibilities of reaching a peace treaty.

June 4, 1967: The Six Day Arab-Israeli War; Deepening Hostilities

Opportunities for compromise between Israel and Syria were virtually nonexistent in the years preceding the Six Day War of 1967. On the Syrian side, a new radical Ba'thist regime emerged in 1963. The Ba'thist regime was committed to an activist anti-Israeli foreign policy for internal and external reasons. Domestically, the Ba'thist regime channeled potential political opposition toward Israel by engaging in a public campaign of extreme anti-Zionist/anti-Jewish propaganda. The Ba'thists accused Israel of joining a Western "colonialist-imperialist" alliance whose objective was to destroy Arab nationalism and expand its military and territorial control over Arab land. On that basis, Israel had to be destroyed by military force. In terms of its foreign policy, the Ba'thists argued that Syria had to increase its economic, political, and military strength before taking up arms against Israel (Ma'oz 1995: 88–89).

In the aftermath of an intraparty coup in 1966, a more militant Ba'thist government emerged under the leadership of senior military officials: Salah Jadid and Hafiz al-Asad. As border clashes over the control of the demilitarized zones near Lake Tiberias increased, the new Syrian government, frustrated with the lack of containing Israeli aggression through the UN Security Council, decided to take stronger retaliatory measures against Israeli actions. In addition to confronting Israel in their border areas, the Syrians embarked on a policy of "liberating" Palestine by extending military bases, training, and arms to the Fatah organization and encouraging Fatah militants to strike against Israeli targets. Their aim was to support a long term "popular liberation war" against Israel along the lines of Vietnam and Algeria. Between February 1966 and May 1967, Syria initiated 177 border incidents and 75 Palestinian guerrilla raids inside Israel.[4]

On the Israeli side, the 1950s and the early 1960s were dominated by Israeli efforts to control the demilitarized zones and divert water from the Jordan River south to the Negev Desert. The irrigation program was perceived to be important for Israel's ability to absorb future immigrants but was an anathema to the Syrians, who perceived it to be a theft of Arab resources that would ultimately lead to Israel's growth at the expense of Palestinian refugees who wanted to return home. Through a series of provocative military raids and clashes throughout the 1950s, the Israelis had by the early 1960s (Seale 1988: 121) captured significant areas, drained the Hula marshes, and won exclusive control of Lake Tiberias; by 1964 Israel had completed its National Water Carrier project (the Jordan River diversion).

In addition to Israel's determination to control the demilitarized zones along Syria's borders, Prime Minister Ben-Gurion and Defense Minister Moshe Dayan also had designs on the Golan Heights for two important reasons. Controlling the Golan Heights would ensure Israel's command of the three important water sources: the Jordan River, the Litani River (in southern Lebanon), and the Yarmuk River (a tributary of the Jordan River in Syria near Jordan). These rivers would guarantee Israel's success in expanding settlements in Palestine and aid its future economic independence. Beside the settlement issue, the Golan Heights also provided greater security against Arab infiltration and shelling of Israeli settlements. Hence, Slater recounts (2002: 90), Ben-Gurion and Dayan supported offensive actions against the Syrians in hopes of widening the conflict so that Israel could seize the area.[5]

During the course of the Six Day War in June 1967, Israel took over the Golan Heights along with the Sinai, Gaza, East Jerusalem, and the West Bank of Jordan. When the war was over, the Israeli government offered the Syrians a peace treaty on the basis of demilitarization of the Golan Heights and a guarantee of a free water flow from the Jordan River into Israel. The Syrians, however, refused, maintaining a policy of "no peace with Israel, no recognition of Israel, no negotiations with it" (see Ma'oz 2002: 213). When the Syrians declined, Israel withdrew the offer and began expelling 130,000 Syrian villagers from the Golan and replacing them with Israeli settlers (Slater 2002: 92). Later, in 1968, the Israeli government formally accepted UN Resolution 242, which called for Israeli withdrawal in the context of permanent peace with secure and recognized boundaries by all parties.

The Syrians, however, rejected the UN resolution and reiterated their goal of liquidating the Israeli state. The military defeat in the war, the loss of the Golan Heights, and the presence of Israeli troops within forty miles of Damascus only hardened the Syrian position toward Israel. Moreover, the war generated an internal struggle for power between the camps of Jadid and Asad over the state's priorities. While Jadid and his cohorts affirmed their commitment to Palestinian guerrillas and a socialist revolution at home, Asad emphasized the necessity of gaining greater control over Palestinian guerrilla actions, achieving military equality with Israel, and bringing an end to Syria's regional and international isolation. By 1970, Asad had defeated Jadid and consolidated his political dominance in the government with the help of the armed forces.

1973: Arab-Israeli War; More Intransigence

At the outset of his regime, Asad was consumed with three problems: 1) redressing the military imbalance between Israeli and Syrian armed forces; 2) reworking Syria's relationship with the Soviet Union; and 3) developing an alliance with Egypt in preparation for another war. As an ardent nationalist, Asad believed that war was the only solution to gaining back the Golan Heights and Palestine. With strong public backing, Asad proceeded to prepare for war. His efforts centered on strengthening Syria's military capabilities and to that end he redirected a large portion of Syria's annual budget to his military. Ma'oz (1995: 122) finds that Asad channeled over 70 percent of the 1971 budget to the army.

Meanwhile, Asad sought a stronger, closer relationship with the Soviets, who at that time perceived the Syrians to be "hotheads" and primarily responsible for the disastrous outcome in the Six Day War (Seale 1988: 148). His success in developing that relationship produced a significant transfer of arms to Syria between 1971 and 1973.

On the eve of the war of October 1973, Syria had 2,000 Soviet tanks and 330 Soviet combat planes (compared to 430 tanks and 150 planes in 1968). It also had 135,000 troops (compared to 60,500 in 1963). These increases were due to backing from the Soviet Union as well as a substantial increase of Syrian military expenditure (from $268 million in 1963 to $384 million in 1970 to $427 million in 1971) (Ma'oz 1995: 122).

Next, Asad secured a military alliance that would force the Israelis to fight a war on two fronts, a perceived prerequisite for success. By the end of 1971, Asad and Egypt's Sadat had reached a secret agreement to launch an attack on Israel with the knowledge and endorsement of the Soviet Union (Seale 1988: 191–193). Meanwhile, between 1971 and 1973, Syria continued to support PLO guerrilla operations against Israel.

On the Israeli side, intransigence persisted through Golda Meir's administration from 1967 to 1974. Meir adopted two fundamental principles that guided Israeli foreign policy after 1967: no return to the prewar border and no withdrawal without direct negotiations and peace treaties with Arab states. So long as the Arabs rejected peace, Israel would consolidate its position along the ceasefire lines, and Israel would not agree to demilitarization policies or international guarantees as substitutes for peace treaties. At the same time, Meir's cabinet was dominated disproportionately by hawks who had a tendency to subordinate diplomacy to military strategic considerations. Consequently, there were substantial internal constraints on Israel's ability to pursue a diplomatic strategy.[6] Finally, according to Shlaim (2001a: 323), more than any other Israeli leader, Meir exhibited a "siege mentality"—a view that Israel had to barricade itself behind an iron wall in the belief that Israelis were doomed to live by the sword.[7]

Egypt and Syria launched a surprise military attack against Israel in October 1973, inflicting serious damage on Israeli troops. More than six thousand Israeli soldiers had been killed or wounded by the time that Israel was able to counterattack successfully and defeat the Egyptians and Syrians. In their war against the Syrians, the Israelis were able eventually to force the Syrians out of new territories that they had conquered and even mobilize troops within twelve miles of Damascus. Nonetheless, at the end of the conflict, the cost to Israel was quite high in terms of lost equipment, and declining economic productivity and exports that totaled $7 billion, equivalent to one year of Israel's gross national product (U.S. Library of Congress 1988).

In the aftermath of the war, the United States brokered a disengagement agreement that was signed in May 1974. The agreement stipulated that Israel would withdraw from new territory that it had captured in 1973 as well as the town of Kuneitra and a small adjacent area. While Syria regained Kuneitra, the best-known site in the Golan Heights, Israel kept the bulk of the area, including Mt. Hermon, a valuable monitoring station (Rabinovich 1998: 28).

The October war in 1973 did not ameliorate tensions between Israel and Syria despite the growing rapprochement between Egypt and Israel that would ultimately culminate in the Camp David Accords of 1979. Although Yitzhak Rabin replaced Golda Meir as prime minister in 1974, the new leadership did not produce a change in Israel's policies toward Syria. While Rabin was ready to negotiate peace settlements with Egypt and Jordan, he was unwilling to make any agreement with the Syrians that would involve an Israeli withdrawal from territory gained in the Six Day War of 1967. Rabin still considered Syria to be Israel's most dangerous and inflexible rival, justifying Israel's continued hold on the Golan Heights as a vital national security requirement. Israeli public opinion, moreover, believed that Syria had launched the October War not only to gain back the Golan Heights but also to grab northern Israeli territory and if possible eliminate Israel (Ma'oz 1995: 141). Hence, Israeli leaders and their public were adamantly opposed to withdrawal even if Syria was prepared to sign a peace treaty.

As for Syria, Asad said that he would sign a nonbelligerency agreement with Israel (not a final peace treaty) in exchange for Israel's withdrawal to the 1967 borders and sovereignty for the Palestinians in the Gaza Strip and the West Bank. In 1974, he announced that he accepted UN Security Resolutions 242 and 338, which called for a comprehensive peace settlement based on Israel's military withdrawal from territory captured in 1967 in return for a guaranteed peace.

Although Asad was interested in pursuing a diplomatic track, the Israelis were not interested. Consequently, he embarked on a policy of attaining military "strategic parity" with the Israelis.[8] His aims were to deter future Israeli aggression in Syria as well as defend Damascus in the event that Israel was not deterred, use his armed forces to launch a small or full-scale war against Israel in order to regain the Golan Heights, and negotiate a comprehensive political settlement with Israel from a position of military strength (Ma'oz 2002: 214).

In an effort to rebuild new Middle East alliances that would supplant the Syrian-Egyptian one, Asad tried to build a confederation with Lebanon, Jordan, and the Palestinians. By 1980, however, Jordan's relations with Syria had soured over their differences concerning Sadat's peace agreement with the Israelis in 1979. Whereas Syria was vehemently opposed to any bilateral negotiations with the Israelis, Jordan quietly backed Sadat's peace initiative in the hope that any peace agreements would lead to similar progress on settling the West Bank–Palestinian issue in line with Jordan's interests (Ma'oz 1995: 156–157).

1981–1985: The Golan Heights and Civil War in Lebanon

Israeli-Syrian relations took another downturn when Menachem Begin emerged as Israel's new prime minister in 1977. Begin and his Likud Party victory brought an end to three decades of Labor rule in Israel, and also brought to power a highly ideological right-wing government. While his first term was consumed with finalizing a peace settlement with Sadat, Begin was less compromising to other Arabs after his reelection in June 1981. In December 1981 Begin, with the support of the Knesset, annexed the Golan Heights, extending Israeli law, jurisdiction, and civil administration to an area that had been under military occupation since 1967. Begin argued that Syria's sustained hostility to Israel and its right to exist justified the annexation. Begin's policy violated "the principles of international law, of Resolution 242, of the Israeli-Syrian disengagement of forces agreement in 1974, and of the Camp David Accords" (Shlaim 2001a: 393). The annexation also foreclosed opportunities for a negotiated settlement with Syria, a policy that every Israeli government since 1967 had avoided. Why did Begin annex the Golan? Shlaim (2001a: 393–394) believes that two reasons stand out: first, Begin wanted to pacify his right-wing constituencies; and second, he wanted to send a message that there would be no further territorial withdrawals after the completion of Israel's withdrawal from the Sinai (as part of the Camp David Accords with Egypt).

Meanwhile, Asad became more concerned about events unfolding in the Lebanese Civil War, which began in 1975. Asad was worried that in the future Israel might attack Syria through Lebanon, or it might capitalize on the instability in Lebanon by expanding its influence in the area—at the expense of Syria's traditional dominance. Asad's fears proved correct when Israel's support of the Christian Maronites—one faction aligned against the Lebanese socialists and the PLO—became public in 1976. At this point, Asad perceived two possible outcomes: the Maronites would create a separate state with Israel as its protector or the radicals supported by the PLO would score a victory over the Maronites, provoking Israel to intervene. If Syria sat on the sidelines too long while the war continued, then Israel was more likely to become involved. Eventually, the Syrians intervened in June 1976 in order to put an end to the civil war but only under the auspices of a U.S. formula referred to as the "red line" agreement between Israel and Syria. This informal agreement established that Syria would not deploy forces in south Lebanon, or use its air force in ground attacks or deploy surface-to-air missiles (SAMs) in Lebanon.[9] Much to Asad's chagrin (Seale 1988: 278–289), Israel capitalized on the new arrangement by funneling arms to its Maronite allies throughout Lebanon and supporting a pro-Israeli militia, led by a former Christian officer of the Lebanese army along the southern border. At the same time, Syria maintained a military presence in Lebanon for the next several years.

The Lebanon War and Its Aftermath

Shortly after his announcement to his cabinet that Israel was going to annex the Golan Heights, Menachem Begin revealed his plans for war in Lebanon on December 20, 1981. Although the architect of the invasion plan was his defense minister, Ariel Sharon, Begin supported it completely. Begin and Sharon's war was designed to accomplish three objectives: 1) to destroy the PLO's military infrastructure in Lebanon and to undermine the PLO's political organization; 2) to establish a new political order in Lebanon by helping the Maronites form a political government that would eventually sign a peace treaty with Israel, and 3) to either expel or substantially weaken Syria's military forces in Lebanon. In the short term, Israel's military invasion would destroy the PLO, undermine Palestinian nationalism in the West Bank, and facilitate Israel's absorption of the West Bank into Israel (Shlaim 2001a: 296).

The Israelis launched "Operation Peace for Galilee" in June 1982. Syrian forces suffered another military defeat by the Israelis. After the United States brokered a ceasefire between Syria and Israel in central Lebanon, Israeli forces moved north to Beirut in order to confront and ultimately defeat the PLO. When the military operations were over in August 1982, all Syrian troops and PLO fighters were evacuated from Beirut during a ceasefire that was mediated by the United States, and shortly afterward a pro-Israeli Maronite leader, Bashir Gemayel, was elected as Lebanon's new president. During the course of the war, the Syrian army suffered severe losses: 400 soldiers killed, 1,400 wounded, 145 tanks destroyed, and 100 planes shot down. In addition, Syria faced a future Israeli satellite in Lebanon (Ma'oz 1995: 176–177).

Facing these major foreign policy setbacks, Asad fought back. First, he sponsored the assassination of Bashir Gemayel, the newly elected Maronite president of Lebanon. Next, when Israeli forces withdrew to southern Lebanon after getting caught between warring Lebanese factions, Asad supported various Lebanese and Palestinian attacks against Israeli troops (as well as U.S. and French peacekeeping soldiers) in an effort to drive them out of Lebanon so that Syria could reestablish its control over Beirut. Syria also sponsored a series of suicide attacks by Hezbollah (a pro-Iranian Shiite militant organization): on April 18, 1983, the U.S. embassy was attacked by a car bomb, killing 63 people; on October 23, 1983, U.S. and French military barracks were destroyed by Shiite suicide squads, killing 241 U.S. and 58 French soldiers; and on November 4, 1983, a similar suicide bombing was launched against Israeli military headquarters in Tyre, Lebanon, killing 28 people (Ma'oz 1995: 178).

Afterward, Asad was able to reestablish his influence over Lebanon when U.S. and French troops left in 1984. At this point, Asad sponsored terrorist attacks by Shiite militant organizations against Israeli forces in southern Lebanon. By 1985, the Israeli government under Shimon Peres had evacuated the Israeli army from Lebanon, leaving a six-mile-wide security belt under the control of a pro-Israeli militia.

The Israeli-Syrian struggle over Lebanon from 1978 to 1985 ended in a major foreign policy setback for Israel. Israel lost its political influence in Lebanon and faced new challenges posed by radicalized anti-Israeli Shiite organizations in southern Lebanon and the rise of the intifada in the West Bank and the Gaza Strip. For Syria, Asad was able to establish his influence over Lebanon and strengthen his relationship with Iran in his efforts to employ militant Shiite organizations against Israel. At the same time, he lost his control over the mainstream Palestinian organizations in Lebanon, Jordan, the West Bank, and the Gaza Strip.

1988–1992: Syria Shifts from a Hard-line Policy; Israel Stonewalls

Between 1988 and 1992, Asad pursued a diplomatic solution to end the impasse with Israel. Announcing his support for UN Resolutions 242 and 338 and his willingness to enter into a peace process, Asad reestablished full diplomatic relations with Egypt, the first Arab country to solidify a peace agreement with the Israelis, as a first step in demonstrating Syria's new flexibility to the United States. In 1990, he expressed his willingness to negotiate directly with Israel in the context of an international conference held by the United States and the Soviet Union and that he was prepared to demilitarize the Golan Heights after it had been returned to Syria.[10] At the end of 1990, Asad met with President George Bush in Geneva—his first visit with a U.S. president in thirteen years.

Meanwhile, in a surprising move in 1990–1991, Asad endorsed a U.S.-led military coalition against Saddam Hussein in the Persian Gulf War, even deploying Syrian soldiers to serve with U.S., European, and Arab troops in Kuwait. After the Persian Gulf War (1991), Asad supported a U.S. plan to attend a peace conference in Madrid where he would negotiate directly with Israel. Two major concessions were made as a consequence. First, Asad agreed to attend a conference that was sponsored largely by the United States and the Soviet Union instead of his first preference for an event sponsored by the United Nations. Next, he dropped a key demand that the Israelis had to commit in advance to a military withdrawal from the occupied territories (in Golan Heights, southern Lebanon, the Gaza Strip, and the West Bank) and to agree to Palestinian statehood. He did secure a U.S. pledge that UN Resolutions 242 and 338 were the legal basis of the peace process and that the United States did not recognize Israel's annexation of the Golan Heights (Hinnebusch 1996: 49).

Despite these concessions, Asad adhered to his long-term position that in return for an Israeli military withdrawal from the occupied territories, Syria would sign a peace treaty, but one that would not include full recognition and formalization of political and economic relations with Israel.

Nonetheless, diplomats and scholars agree that Syria's foreign policy shifts after 1988 signaled a new direction to the problem of regaining the Golan Heights and solving the Palestinian problem. Relying on U.S. aid also reflected a change in Syria's past policies, which had rejected U.S. strategies that supported direct negotiations with Israel (Cobban 1999; Ma'oz 1995, 2002; Rabinovich 1998).

What accounts for Asad's new flexibility and concessions toward Israel? Two major reasons stand out. First, the global distribution of power had changed dramatically by the early 1990s with the collapse of the Soviet Union, the breakup of the Eastern European bloc, and the rise of the United States as the sole superpower. Syria had relied heavily on the Soviet Union and the East European bloc to finance much of its domestic economic and foreign policies. Starting in the mid-1980s, the Soviets reduced their military assistance to the Syrians, encouraging Asad to pursue a diplomatic solution to a peace settlement with Israel.[11] In addition, the Soviets opened new lines of dialogue with the Israeli government, permitting Soviet-Jewish emigrations to Israel. Finally, as the Cold War came to an end in the 1990s, Asad could no longer manipulate superpower rivalries to his advantage. As the Soviets chose to play a junior partner role to the United States in trying to resolve the region's problems, Drysdale and Hinnebusch (1991) argue that Asad came to the conclusion that he could no longer achieve his goals without U.S. support.[12]

A second related reason is the rise in U.S. prestige and influence in the Middle East after the Persian Gulf War of 1991, which set the stage for an aggressive search by the Bush administration for a comprehensive solution to the Arab-Israeli conflict.[13] After nine trips to the Middle East, the U.S. secretary of state, James Baker, put together the formula for an international conference in Madrid. Along the way, the United States obtained another important concession from the Syrians that allowed a second set of multilateral negotiations to accompany the bilateral talks. The multilateral negotiations focused on five regional problems: water, refugees, arms control and regional security, the environment, and economic cooperation. These negotiations were designed to facilitate concessions that would have to occur among the parties on the bilateral track. The Madrid Conference was the first U.S. effort to bring together both the Palestinians and the Arab states at the same time (Rabinovich 2004: 33–36).

How did the Israelis react to both Syrian peace initiatives and the Madrid Conference? Israel's hard-line prime minister, Yitzhak Shamir, went to

Madrid in what Shlaim (2001a: 488) describes as a "defiant and truculent mood." After handing back the Sinai to Egypt, Shamir's right-wing government refused to make further territorial concessions to Syria or the Palestinians. Shamir rejected Syria's conditions for a comprehensive peace and in particular a withdrawal from the Golan Heights; he also argued that UN Resolution 242 did not apply to the Golan Heights. After withdrawing from the Sinai in 1979, Shamir was ideologically opposed to handing back any more territories to the Arabs. According to Shlaim (2001a: 501), Shamir believed that permanent conflict existed between the Israelis and the Arabs; that Israel would always be surrounded by enemies who were "untrustworthy, inherently vicious and unalterably committed to its destruction." Not long afterward, in June 1992, the Israeli electorate voted Shamir and the Likud Party out of office in favor of Labor's candidate, Yitzhak Rabin, in a landslide victory. At this point, Israel significantly changed its negotiating position with the Syrians.

1992: Rabin Emerges as a Policy Entrepreneur

After his election, Rabin was determined to change the national agenda by revitalizing the Madrid peace process. Unlike his predecessors (especially Shamir), Rabin perceived that he had a small window of opportunity to achieve serious breakthroughs for peace with his Arab rivals. He maintained that significant changes at the international and regional levels made it possible for Israel to change its hard-line foreign policy. For instance, the Soviet Union's collapse was a significant, positive change, creating more possibilities for Middle East peace, primarily because Syria and the Palestinians had lost an important ally that had earlier aided their ability to confront Israel. At the same time, the rise of the United States, Israel's close ally, to hegemonic status was a stabilizing force for the Middle East peace process. Another change involved Israel's new international status. Instead of seeing Israel as an isolated country whose security depended primarily on its own resources, the international community was less hostile toward Israel, so Rabin now believed. In his inaugural address, Rabin spoke of an emerging "new world" that was based on "peace, reconciliation and international cooperation." He argued that it was Israel's duty to explore the risks as well as the opportunities for making peace (Inbar 1999: 134–135).

At the regional level, Rabin identified several other developments that augured well for Israel. One such development was Iraq's defeat in the Persian Gulf War of 1991. Not only was Iraq, a major rival, defeated, but the U.S. protection of Israel (through extended deterrence) would discourage other Arab military challenges in the future. Another development was the immigration of over half a million people from the former Soviet Union to Israel. The influx of so many people created a significant demographic shift that other Arab rivals could not eliminate. Hence, the threat to Israel's existence from Syria had decreased substantially. At the same time, Rabin perceived new threats from Iran (and its efforts to acquire a nuclear weapon), and Islamic extremism in the form of terrorism. In fact, Inbar (1999: 135–136) argues that Rabin now believed that Iran was Israel's primary threat.

Rabin also believed that the timing for pursuing peace in 1992 was propitious. Israel was now able to negotiate from a position of strength considering Syria's obvious military weaknesses and its recent loss of an important ally, the former Soviet Union. Moreover, these new threats from Iran and Islamic radicals made it imperative to move quickly on negotiating a peace settlement with Syria (Inbar 1999: 140).

Another factor thought to have increased Rabin's sense of urgency (Inbar 1999: 140) was the fragile domestic political situation. Rabin was worried about how long he could hold together the slim parliamentary majority that supported his government. Thus, he wanted to expedite the negotiations, which he did, with the Palestinians, who indicated that they were willing to make major concessions. Dealing with the Syrians, however, was going to be more difficult.

Historically, Rabin and the Labor Party had insisted that any peace agreement with the Syrians would always entail continued Israeli settlements and an Israeli military presence in the Golan Heights. In fact, the Labor Party over the years had encouraged Israeli settlers to move to the Golan. So, any compromise with the Syrians on the basis of a military withdrawal would be very controversial and risky domestically since considerable opposition was likely to emerge from elements within his own party as well as the Likud and other right-wing parties (Cobban 1999: 71; Ma'oz 1995, 221–222).

Yet Rabin knew that the Syrians were adamantly opposed to any settlement that did not involve total Israeli military withdrawal from territory gained in 1967. Just a few months after Rabin's election, the Syrian foreign minister, Faruq al-Shara, had expressed Syria's willingness to sign a final status agreement (total peace) in return for Israeli military withdrawal from Syrian land acquired in 1967. By a final agreement, the Syrians meant more than a nonbelligerency agreement; they were now willing to establish full diplomatic and economic relations with Israel but only in the context of a full military withdrawal. Consequently, Rabin was doubtful that the positions between Syria and Israel could be reconciled.

Rabin changed his perspective after meeting with James Baker, the U.S. secretary of state, who, after traveling to meet with Asad in Damascus in July 1992, conveyed his impression that Asad was serious about making peace with Israel and that the United States would make a significant investment in bringing it about.[14] While Rabin thought that making peace with the Palestinians would be faster and easier, he decided to pursue separate negotiations with both the Syrians and the Palestinians secretly (Rabinovich 2004: 44).

In terms of a deal with the Syrians, four major issues had to be confronted: 1) the return of the Golan Heights under conditions that would not threaten Israeli security; 2) disposition of water resources on the Golan; 3) the precise location of the Syrian-Israeli border; and 4) terms and timing of the normalization process involving economic and political cooperation. Over the next several years of negotiation, Syria and Israel settled on a draft agreement that resolved most of these differences. Both sides made major concessions. Namely, over a period of three to five years, Israel would withdraw from the Golan Heights to the border of June 4, 1967 and full Syrian sovereignty would be restored. In return, Syria would agree to the demilitarization of the Golan Heights, a water-sharing agreement with the Israelis, and full normalization of relations corresponding to Israel's staged withdrawals.[15]

The most significant Israeli concession involved Rabin's agreement to a military withdrawal to the borderline of June 4, 1967 in exchange for full normalization of relations with Syria.[16] Rabin discarded Israel's traditional position: that military withdrawal would be made at the international boundaries negotiated between France and Britain in 1923. This compromise meant that the Syrians would once again have a position on Lake Tiberias. Between 1992 and 1994, Israel and Syria had settled most of their outstanding differences and were close to a final agreement. By the end of 1995, however, the momentum toward a final settlement had slowed considerably and it eventually died out altogether in May 1996 when the Israeli public elected a new right-wing government under Binyamin Netanyahu's leadership.

1993–2000: Deceleration and Eventual Stalemate in Syrian-Israeli Negotiations

YITZHAK RABIN

Between 1993 and 1995, the pace of Syrian-Israeli negotiations slowed considerably for three reasons. One reason is that Rabin zigzagged between giving priority to the Syrian and to the Palestinian tracks, playing one party against the other (Cobban 1999: 53; Inbar 1999: 143). As he engaged the Syrians and Palestinians in separate bilateral talks, Rabin used the negotiations to blunt Asad's attack on the peace talks with the Palestinians (at Oslo), while pressuring the Palestinians into making major concessions. He was aware that the Palestinians were "terrified" at the prospect of Israel signing a deal with the Syrians first, leaving them alone and more vulnerable than ever (Cobban 1999: 51; Seale 2000: 68). Consequently, the PLO negotiated the Oslo Accords in September 1993, agreeing to end the intifada in exchange for its political recognition. The remaining issues dealing with settlements, refugees, borders, and Jerusalem would be addressed later. Meanwhile, Rabin, knowing that the Syrians would be unwilling to make any territorial concessions on the Golan Heights, hesitated to move on the Syrian track, choosing the Palestinian one instead (Inbar 1999: 144).

A second reason for the slowdown in negotiations between Israel and Syria is that Rabin purposely delayed the process by pursuing the Jordanian peace process. Realizing that the Syrians were not going to approve a "vague, piece-meal and open-ended agreement like the Oslo Accords" (Hinnebusch 1996: 55), Rabin turned his attention to the Jordanians and subordinated the Syrian track.[17] In addition, the Jordanian negotiations were more appealing because Rabin continued to worry about the lack of domestic support for Israeli withdrawal from the Golan.[18] He also argued that the Israeli public needed time to adjust to the Oslo Accords.

Later, Asad became convinced that Rabin was deliberately wasting time, particularly when the U.S. government put little pressure on Rabin to move forward in the negotiations (Seale 2000: 74). Meanwhile, in January 1994, Rabin added another condition to the peace process; namely, that any peace agreement with the Syrians had to be put to a public referendum. The purpose of the referendum was to neutralize domestic opposition and prepare Israelis for a significant withdrawal on the Golan. Ma'oz (1995: 244) believes that Rabin also meant to "throw the ball back into Asad's court" in order to pressure him to make public peace gestures (similar to Sadat's visit to Israel in 1977).

Rabin hoped that such overtures would shore up domestic political support.[19] In fact, advisers close to Rabin have disclosed that since the summer of 1993 Rabin had already made up his mind that any significant breakthroughs with Syria would have to wait until his second electoral term. "Hesitant and indecisive," Rabin did not believe that he had a public mandate to withdraw from the Golan Heights since he had not raised the issue during his electoral campaign in 1992. He also did not want to confront settlers in the Golan, many of whom were supporters of the Labor Party. Eventually, he came to believe that concluding a deal with the Syrians prior to the national election of 1996 would cost him valuable votes and decrease his chances of winning reelection. Therefore, Rabin postponed a breakthrough until his second term when he would be stronger and hopefully Asad would be weaker (Inbar 1999: 147–148; Seale 2000: 75).

The last reason for the loss of momentum concerns the opposing views that Syrians and Israelis had on what the peace process would eventually produce. For Asad, who as defense minister in 1967 lost the Golan Heights during the war of that year, complete return of the territory was a symbol of Syrian sovereignty and independence. Equally important, though, was the security issue. The Golan represented a defensive buffer against Israel, whose military presence in the area constituted a serious threat. On a broader level, the Syrians hoped to use a peace agreement to "rollback and contain" Israeli influence within its pre-1967 lines, thereby preventing Israel's absorption of Lebanon, Jordan, and the Palestinians into its sphere of influence at the expense of Syria's regional ambitions (Hinnebusch 1996: 51).

For the Israelis, a peace agreement would result in full normalization of relations, including diplomatic ties, trade, and tourism. In sum, the agreement would permit its incorporation into the Middle East (Hinnebusch 1996: 51). In contrast, the Syrians were not willing to give any more advantages to the Israelis, whom they believed sought regional supremacy under the cover of a peace process (Seale 2000: 71).

Throughout 1995, Rabin continued to seek public assurances from Asad that Syria was truly ready to sign a peace treaty with Israel. As Asad rejected Rabin's request for summit meetings or other forms of confidence-building measures, Rabin believed that he could do no more to bring about an Israeli-Syrian peace. In November 1995, Rabin was assassinated by a right-wing Jewish law student. Shortly afterward, Shimon Peres, who was serving as Rabin's foreign minister, was named prime minister.

SHIMON PERES

Unlike Rabin, who was skeptical about the prospects of peace, Shimon Peres had long sought a comprehensive Middle East peace with Israel's Arab neighbors. He believed that it was wrong of Israel to perpetuate the territorial status quo and to rest its security only on massive and costly armed forces. He advocated Israeli withdrawal from the occupied territories, conflict resolution, and open borders that would integrate Israel economically throughout the Middle East, from North Africa to the Persian Gulf. He was a strong proponent of the political benefits that could be derived from economic integration. When Peres joined Rabin's government, he made it clear that his primary goal was making peace with Israel's neighbors (Shlaim 2001a: 505–506).

Whereas Rabin had discussed security issues with the Syrians, Peres was more concerned with issues of regional economic development and water security. In particular, Peres wanted guarantees that Syria would not hamper the flow of water from the Golan to Lake Tiberias, and he supported joint economic ventures in developing the area. Asad was willing to make promises about the water issues, but he balked at the joint economic cooperation, arguing that the Syrians would see such cooperation as an extension of the occupation and reflect Israeli efforts to establish hegemony (Seale 2000: 76).

Despite these differences, the Syrian-Israeli talks resumed more intensely at the Wye Plantation conference center near Washington, D.C. in December 1995. Several more rounds of talks occurred in January and February 1996. By the end of February 1996, the Syrians and Israelis had completed about 75 percent of the negotiations toward an agreement and were hammering out the remaining details on a timetable for military withdrawal and the process of normalization. Yet, in March 1996, Peres suspended the Wye talks and brought four and a half years of Israeli-Syrian talks to a halt (Cobban 1999: 136).

Peres's reason for suspending the talks had more to do with his decision on the timing of the next national election. Initially, Peres was interested in a later date (October 1996) with the aim of concluding a peace treaty with the Syrians that would propel him into an electoral victory. As a consequence, Peres began to push for a speedy conclusion to the talks, requesting a summit meeting with Asad to push through the final details. Asad, in contrast, found it premature to meet when there were still remaining gaps between the Israeli and Syrian positions that precluded a final agreement (Cobban 1999: 142; Shlaim 2001a: 554). When he realized that the process was unlikely to yield results by October 1996, Peres decided (in February) to hold an early election in May 1996, calculating that he could a win mandate for another four years and then renew his efforts to finalize a peace treaty with Asad (Shlaim 2001a: 554).

Instead Peres lost the national election. A series of suicide bombing attacks from Palestinian radicals in early March, killing sixty Israeli citizens, reverberated negatively on Peres's government. Public opinion polls tilted in the direction of Binyamin Netanyahu, a hard-line leader of the Likud Party, while thousands of right-wing supporters demonstrated against Peres and the peace process. In an effort to protect himself from political attack during the preelection period, Peres suspended the talks with the Syrians. Approximately a month later (April 1996), he launched a small military intervention (Operation Grapes of Wrath) against Hezbollah forces in southern Lebanon.[20] By the end of the month, the operation was over, but it was largely unsuccessful in eradicating Hezbollah rocket attacks into Israel. Meanwhile, Peres lost critical votes from Israel's Arab voters who were highly critical of the operation, and among Jewish voters who perceived that Peres was incapable of ensuring their personal safety. Peres garnered 49.6 percent of the total vote to Netanyahu's 50.4 percent (Shlaim 2001a: 559–563).

BINYAMIN NETANYAHU

Netanyahu presented an alternative vision of peace with Syria that essentially repudiated his predecessors' policies. He argued that in the absence of democratic regimes in the Arab world, Israel could not expect a genuine peace; that it could only be attained through military strength and deterrence (Rabinovich 1998: 263). Moreover, his image of the Arabs was consistently negative and did not consider the possibility of diversity or change (Shlaim 2001a 566).

Consequently, he felt no obligation to adhere to the policies of Rabin and Peres vis-à-vis the Syrians. He made it clear that he would not accept the basis of the Madrid peace process—the return of land in exchange for peace.

In short, any peace treaty with the Syrians would not involve a military withdrawal from the Golan Heights regardless of the commitments made earlier by both Rabin and Peres about Israel's withdrawal to the borders of June 4, 1967 in the context of a final peace agreement. In the end, the Israeli government remained decidedly against resuming peace talks with the Syrians for the next three years until a new Labor-led government under Prime Minister Ehud Barak emerged in May 1999.

EHUD BARAK

After Barak's electoral victory over Netanyahu in May 1999, two rounds of Israeli-Syrian negotiations occurred in December 1999 and January 2000 under the auspices of the Clinton administration in Washington, D.C. and later Shepherdstown, West Virginia. Outstanding issues remained on the timetable for military withdrawal, the degree of demilitarization along the border area, and guarantees about Israel's continued access to the Jordan River, its tributaries, and Lake Tiberias. However, the issue about Israel's military withdrawal to the border of June 1967 emerged as the key obstacle. Even though Barak had recognized that Rabin and Peres had committed Israel to this policy, he was unwilling to negotiate the final details, primarily due to the absence of domestic support and some form of public diplomacy from Asad that would have made it easier to sell the deal to the Israeli public (Slater 2002: 98–100). After the talks broke down in Shepherdstown, two months followed without any progress. During this time, Asad's ability to advance the talks was hampered by his failing health, and his preoccupation on securing his son's succession at a time when Israeli domestic criticism over the peace talks was increasing. By March 2000, attempts to revive the negotiations had once more been precluded by two developments: Israel's withdrawal from Lebanon in May and Asad's death in June (Rabinovich 2004: 130–140).

Implications for the Model

Of all of the rivalries in the Arab-Israeli conflict, the Israeli-Syrian rivalry has been the most intense and difficult to resolve. Mirror images of Israeli and Syrian distrust and hatred have dominated the political discourse between these governments and their societies. Both governments have engaged in "excessive demonization and propagandizing, to the point that their populations have virtually no grip on facts nor an informed understanding of the enemy" (Kessler Neff 2000: 69). Therefore, the breakthrough toward negotiations in the 1990s is remarkable and primarily the result of the leadership of Asad, Rabin, Peres, and Barak. On the Syrian side, Asad pursued the peace process when, in the presence of repeated military defeats at the hands of the Israelis and the loss of Soviet patronage at the end of the Cold War, he realized that Syria would probably never attain strategic parity with Israel, diminishing its ability to compete over the long term. On the Israeli side, Rabin initiated talks when he perceived declining threats posed by Israel's traditional rivals and the emergence of new threats from Iran and Islamic fundamentalists. Moreover, the end of Israel's isolation in a changing global environment (in the aftermath of the Cold War and the Soviet Union's collapse) coupled with a close relationship with the world's remaining superpower, the United States, also contributed to Rabin's perception of optimism and flexibility toward securing peace with Syria.

Our model of rivalry de-escalation captures this process quite well, particularly after 1988. Shocks played an important role in shaping the decision-making processes of both the Syrians and the Israelis, although not every shock was associated with dramatic shifts in thinking or future expectations. The fundamental question is why in the midst of these shocks and strong intermittent mediation by a third party (for example, the United States) did the Israeli-Syrian rivalry fail to terminate. The answer is that despite the shocks' effects on shifting expectations between the Israelis and the Syrians and thus de-escalating the rivalry, expectations were not revised significantly enough to bring about a final peace agreement. One reason is that the incentives to alter those expectations were not strong enough.

According to Cobban (1999: 179–180), both Israel and Syria brought important strategic relationships to their negotiations in the 1990s. While Israel enjoyed a close relationship with the United States, and over time better relationships with the Europeans and some Middle Eastern governments, the Syrians had valuable alliances as well; namely, Egypt, Saudi Arabia, Lebanon, the Gulf Cooperation Council states, and Iran. Consequently, neither state was isolated and neither had experienced a recent significant political-military crisis that might have required a rapid rethinking of its policies and forced one or both of the parties to the negotiating table. In short, neither side had experienced a strong political shock that influenced their relationship. Both Israel and Syria also had strong fallback positions should the negotiations fail. For instance, the Israelis would maintain control over the Golan in the advent of failure, while for the Syrians, failure would preserve the status quo without alienating any important allies. Hence, the lack of urgency in the negotiation process encouraged both governments to view the final peace agreement as optional. According to Uri Savir, who headed the Syrian negotiations during Peres's administration, "There was a sense among both delegations that, if necessary, we could go on living without peace."[21]

Another consideration is Rabin's procrastination in sealing a deal with the Syrians. At the Madrid Conference in 1991, the Syrians had already surmounted major psychological and strategic obstacles to get there (Kessler Neff 2000: 70). Israel, in contrast, was always the stronger party, able to dominate the pace, agenda, and shape of the negotiations. Considering these advantages, Rabin neglected to pursue negotiations aggressively with the Syrians because he was more interested in pursuing "divide-and-rule" tactics pitting Syria against Jordan and the PLO, with whom he was also negotiating. According to Cobban (1999: 186), this strategy signaled that Rabin still viewed the relationship with Syria from a "paradigm of competition"—an approach that made sense when Israel was at war with its rival but was not necessarily helpful when it was pursuing regional peace.[22]

Differences in future expectations about what the peace process would eventually yield were also obstacles to rivalry termination. For Rabin, Peres, and Barak, peace would mean more than just security arrangements; it would also produce full diplomatic and economic relations between Israel and Syria, signaling Israel's full incorporation in the Middle East. According to Peres, the normalization process meant that "the New Middle East would be dominated by banks, not tanks, ballots, not bullets and where the only generals would be General Motors and General Electric" (Shlaim 2001a: 552). Over time when there was little progress in the peace talks, Israeli negotiators began to believe that a final peace could not be secured because Asad was more interested in getting back the Golan Heights and building a better relationship with the United States, rather than seeking a genuine peace on its merits alone (Kessler Neff 2000: 82). Moreover, they argued that Syria's concept of peace was based on the tenet that Israel would no longer be an enemy but would remain an "adversary and competitor" and that the struggle with Israel would continue in other nonmilitary ways (Rabinovich 2004: 134).

The Syrians, in contrast, believed that the outcome of the peace process would result in recovery of the Golan Heights, access to Golan water sources, enhanced security as a result of Israeli military withdrawal from the Golan, greater access to the Western-dominated world market, greater private investment capital, and possibly an international aid package (Hinnebusch 1996: 56). Syrian diplomats, however, resisted normalization as a precondition of peace. In their view, they wanted to keep the "peace as cold as possible" while phasing in normalization very slowly. In the Syrian view, open markets and borders would serve to increase Israel's hegemonic control of Arab resources and ultimately diminish Arab nationalism and unity. Finally, for Asad, a peace agreement meant containing Israel, to "shrink its influence to more modest and less aggressive proportions," which Arabs could accept and live with. Asad was determined that Israel should not derive any further strategic, economic, or political advantages from the peace process (Seale and Butler 1996: 36–37).

The model stresses two other important ingredients to rivalry termination: reciprocity and reinforcement. Thompson (2005) maintains that once new strategies are experimented with, intransigence (or the lack of reciprocity by the adversary) and/or the failure of new strategies to achieve a final settlement (the lack of reinforcement) are likely to lead to an abandonment of new approaches and a relapse to traditional strategies. In other words, the new relationship is highly unstable and expectational revisions require constant reinforcement.

The lack of reinforcement is particularly acute in the Israeli-Syrian rivalry de-escalation phase. Rabin, Peres, and Barak all sought some sort of public diplomacy between Israel and Syria. They wanted a series of public gestures to persuade themselves and their electorates that Syria had "had a change of heart and was now ready for genuine peace with yesterday's enemy" (Rabinovich 2004: 132). From the Israeli view, peace negotiations were to be conducted in the midst of confidence-building measures that were to be based on interim steps, phased implementation, and active public diplomacy. Such measures would increase mutual trust and create the conditions for reciprocal concessions (Rabinovich 1998: 101–102).

Asad, however, was adamantly opposed to public diplomacy. He and his advisors believed that it was Israel's responsibility to prepare its public for peace with Syria. In addition, Asad maintained that public gestures, major summit meetings, and other forms of public diplomacy came at the end of the negotiating process when the final peace treaty had been signed. Like most Syrians, Asad believed that public diplomacy and short-term confidence-building measures involved strategies that were designed to commit Syria to partial or interim agreements that Israel could later delay or abort. They perceived that they were resisting Israeli "traps" that could produce similar mistakes that the Palestinians had made in the Wye and Oslo Accords (Kessler Neff 2000: 79).[23] Finally, the Syrians maintained that Israel's unwillingness to prepare its public for peace diminished their confidence that the Israelis were serious in their intentions to finalize a peace agreement (Kessler Neff 2000: 79).

One final consideration is the role of consolidated policy entrepreneurs in bringing about expectational revisions. Indeed, Asad, Rabin, Peres, and Barak can all be considered entrepreneurs who played significant roles in de-escalating the Israeli-Syrian rivalry. However, Israeli public opinion frequently influenced the negotiation process, and in many respects, Rabin, Peres, and Barak were unwilling to seal a final agreement with Syria because of their concerns about their political survival. Although Rabin was a "leader of unrivaled credibility within the Israeli political system on all matters of war and peace," he delayed negotiations with the Syrians long enough to allow domestic political groups to lobby against major concessions (Cobban 1999: 184). His assassination in 1995 slowed the momentum even more. Nonetheless, when Peres took over the government's leadership, he was eager to continue the negotiations with the Syrians. Because he lacked the credibility that had been Rabin's greatest asset, Peres was too insecure to continue the negotiations without securing his political position through a national election, thereby ending the peace talks. Three years later, Barak emerged as the next Israeli leader poised to negotiate the final peace treaty with Syria. Like Peres, Barak got "cold feet" primarily because the public and other domestic groups were against it, and he, too, backed away from the negotiations (Rabinovich 2004: 133). Policy entrepreneurs are indeed important, but in cases where they are weak or have yet to consolidate their domestic political positions securely enough to negotiate controversial agreements, rivalry termination will be almost impossible. Weak political entrepreneurs can experience expectational revisions, and they can move their countries toward de-escalation, but their ability to move to the final steps of finalizing a deal is compromised in light of their limited abilities to fulfill the necessary reciprocal and reinforcement measures of rivalry termination.

Our model has been helpful in understanding both the de-escalation of the Israeli-Syrian rivalry and the absence of rivalry termination in this case. The main theoretical question in this chapter is to assess whether expectations are critical to rivalry de-escalation as the model suggests, whether shocks are needed to tip expectations, and what mix of other factors are important to attaining rivalry termination. The Israeli-Syrian case demonstrates that indeed shocks were crucial to changing expectations, but these shocks did not provide strong enough incentives for the two parties to alter their expectations very deeply. At the same time, the case shows that policy entrepreneurs indeed play an important role in revising expectations, but unless these leaders are truly secure in their domestic political situations, they are unlikely to bring about an end to a rivalry. Finally, the Israeli-Syrian case demonstrates that expectational revisions that are not coupled with reciprocity and reinforcement are unlikely to produce rivalry termination.

The shocks in the Israeli-Syrian rivalry, summarized in Table 5.1, are uniform in the sense that they all fall in the same "shift in competitive ability" column. Three Syrian war defeats, one series of border clashes, and a loss of Soviet patronage are similar in a different sense. All but one were certainly more shocks to the Syrian political system than they were to the Israelis. The exception was the 1973 war, which was regarded by both sides as shocking. While one might say that the Israelis were the principal source of the first three shocks, the fourth one was more symmetrical. The fifth one, the Syrian loss of Soviet patronage, was asymmetrical.

Table 5.1. Shocks in the Israeli-Syrian Rivalry

Time frame of shocks	*Shift in external threat*	*Shift in competitive ability*	*Shift in regime orientations/ strategies*	*Economic resource crisis*
1949		War defeat (Syria)		
1952		Border clashes (Syria)		
1967		War defeat (Syria)		
1974		War defeat (Syria)		
1990–1991	Iran and radical Islamic terrorism	Loss of Soviet patronage (Syria)		
1992–1996				

Yet none of the first five shocks had any apparent effect on the expectancies of either the Syrians or the Israelis. The Syrian initiatives in the first three attempts at negotiating some type of de-escalation reflected tactical considerations. The external antagonist was not seen in any different light. Israel and Syria remained threatening antagonists, but aspiring negotiators thought something might be gained in the short term by reducing tensions. No reciprocity, as registered in Table 5.2, was manifested in any of the first five episodes. Not surprisingly, none of these cases reached the reinforcement stage either. Change entrepreneurs were largely absent. Third-party pressure was rarely a major factor except in the fifth, the 1990–1991 case.

The one noteworthy exception among the six iterations is the 1992–1996 case. No new shock is related directly to this case, but there was nevertheless some reduction in threat perception and, thus, some expectancy revision. Rabin could be classified as a new change entrepreneur, but he was not committed to changing the Syrian relationship as a high priority. Pressure from the United States was evident. Some reciprocity was exchanged but only for a limited time. A change in Israeli governments brought an end to any reinforcement moves from the Israeli side. The ultimate outcome, as a consequence, was an absence of meaningful change in the Israeli-Syrian relationship.

The role of shocks seems greatly exaggerated in this case. The Israeli-Syrian rivalry has experienced quite a few of them, yet the only serious movement toward de-escalation is associated with the absence of any triggering shock. Third-party pressure was at least facilitative but obviously not sufficient since it was present in three cases overtly. Expectancy revision appears to be important. The only one of the six cases in which it is demonstrated did move toward the possibility of de-escalation. One or more change entrepreneurs also seems important for the same reason. The role of this factor, nonetheless, is ambiguous given Rabin's own hesitancy in pursuing a Syrian rapprochement. Finally, the virtual absence of reciprocity and reinforcement suggests that they are significant factors by default.

Many of the factors associated with Israeli-Syrian interactions are also found in the history of Israeli-Palestinian relations. One difference, however, is that Syria has been a state since the beginning of its rivalry with Israel while Palestine has not. It seemed to begin moving toward the attainment of state independence in the late 1980s and the early 1990s. In this respect, the rivalry between Israelis and Palestinians can be treated on lines similar to the ways we treat interstate relations. The last part of this chapter switches the focus to the Israeli-Palestinian relationship and one of the best

Table 5.2. Modeling Outcomes for the Israeli-Syrian Rivalry, 1948–2000

Variable	*1949*	*1952*	*1967*	*1974*	*1990–1991*	*1992–1996*
Shocks	Present	Present	Present	Present	Present	Absent
Expectancy revision	Absent	Absent	Absent	Absent	Absent	Limited
Consolidated and new policy entrepreneur	Absent	Absent	Absent	Absent	Absent	Limited
Third-party pressure	Absent	Absent	Absent	Limited	Present	Present
Reciprocity	Absent	Absent	Absent	Absent	Absent	Limited
Reinforcement	Absent	Absent	Absent	Absent	Absent	Absent
Outcome	No change	No change	No change	No change	No change	No change

opportunities to alter it significantly in the wake of the intifada shock of the late 1980s.

The Israeli-Palestinian Rivalry

One of the problems with the 1979 Camp David Accords de-escalating to some extent the Egyptian-Israeli rivalry is that the accords encompassed a vague agreement to proceed toward a resolution of the Israeli occupation of the West Bank and Gaza in the aftermath of the Six Day War of 1967. Populated primarily by Palestinians, these areas once controlled by Jordan and Egypt were expected to form an autonomous Palestinian territory. However, little was done to operationalize this component of the Camp David Accords and with Egypt no longer likely to threaten a military attack on Israel, Israeli officials felt little pressure to proceed. Underlying the problem were conservative groups in Israel who felt that control of the Palestinian territories should be made permanent as part of a Jewish return to the ancient areas of Judea and Samara. Control of Jerusalem was also a problem in the sense that it was felt difficult to divide it (into eastern and western zones) and yet no one wished to cede control to opposing groups. Intermixed with religious preferences were Israeli security concerns about creating territorial buffers from future attacks.

The goals of right-wing Israeli groups to maximize control of Palestinian territory were, of course, a form of mirror image to the original goal of Arab and Palestinian groups, dating back at least to the internationalized civil war over control of Palestine in 1948–1949 in which Jewish successes in controlling much of the former British mandate had led to the creation of the state of Israel. A Palestinian diaspora had been created then by Arab refugees fleeing Israeli-held territory. In turn, that created the vexing problem of the "right of return" over whether the refugees would ever be allowed to return to their former homes. One way to accomplish this goal involved evicting the Israelis and the Jewish state and creating an Arab state of Palestine. Naturally, this approach was not widely appreciated in the new state of Israel. Nor was there any desire to see the refugees return or to increase the size of the Arab population within Israel.

In the 1950s and 1960s, guerilla attacks on Israeli territory were launched from adjacent Arab states. Gradually, the Palestine Liberation Organization (PLO), established as a coalition of Palestinian groups, developed into the main organizational embodiment of the resistance to the state of Israel. In the same period, however, Israeli retaliation for guerrilla attacks was strong and intended to convince the adjacent Arab states to suppress further attacks by nonstate groups. This process led to a convoluted history of civil wars in Jordan (1970) and Lebanon (1975–1990) over, at least in part, how much operating freedom Palestinian militant groups would have in these states to attack Israeli objectives. Israel almost intervened in 1970 in support of the Jordanian monarchy and did intervene in Lebanon, especially in the war in 1982. The end result was the weakening of the various external Palestinian groups' abilities to inflict damage on Israeli targets.

Internal resistance to continued Israeli control of the West Bank and Gaza flared rather prominently in the intifada or uprising, manifested as riots, demonstrations, and economic boycotts, which began at the end of 1987 and continued through 1991. The PLO eventually gained some degree of control over these resistance activities but, at the same time, was damaged politically and financially by supporting the Iraqi invasion of Kuwait. It had already lost its Soviet support with the collapse of the Soviet Union. Now it also lost its financial support from the wealthy Arab states that had supported the Persian Gulf War coalition to evict Iraq from control of Kuwait. Although not part of the coalition, Israel also came under Iraqi attack in the ensuing war. In general, the defeat of Saddam Hussein in 1991 led to a renewal of major power interest in resolving the Arab-Israeli conflict and, in particular, the long-disputed disposition of the occupied territories of the West Bank and Gaza and the possibility of creating an Arab Palestinian state. The initial effort focused on talks among Arab states and Israel held in Madrid, but the focus shifted to initially secret, more direct negotiations between Israel and the PLO.

In 1993, Yitzhak Rabin and Yasser Arafat, representing Israel and the PLO respectively, signed the Oslo Accords. These accords were to establish a procedure for negotiating the specifics of which parts of the West Bank and Gaza would come under autonomous Palestinian rule. The Israeli position was that not all the occupied territories could be returned to Arab rule due to the many Jewish settlements that had been established in these areas since 1967. In addition to exempting most of the settlements from transfer, there would also be a need for Israeli access to the defense of the settlements, which meant control of roads and military checkpoints. The question subsequently became just how much of the occupied territories would come under the control of a Palestinian authority.

However, Rabin was assassinated in 1995 and the conservative Likud Party came back into power in 1996 with the intention of slowing down the operationalization of the Oslo Accords negotiation process. The Likud Party lost power in 1999 to a coalition government and one more attempt to strike a concrete deal with the PLO was attempted in 2000 at the Camp David Summit. The Israeli offer in 2000 was rejected by Arafat, leading to a second intifada and a serious and continuing challenge to the likelihood of a viable, two-state solution to the future of Palestine.

Whatever the future of this protracted conflict, our question is how was it possible to arrive in 1993 at a possibility of some resolution of the dispute? More specifically, is our emphasis on shocks, expectancy revision, policy entrepreneurs, third-party pressures, and reciprocity/reinforcement helpful in accounting for this opportunity to de-escalate Israeli-Palestinian antagonisms? Table 5.3 provides a quick list of the highlights in the immediate history that preceded the accords of 1993. Within this backdrop, we proceed to a discussion of the role of shocks in disturbing the Israeli-Palestinian inertia.

Palestinian Shocks

THE ISRAELI INVASION OF LEBANON IN 1982 AND THE REBELLION WITHIN THE PLO IN 1983

The PLO's defeat in the Lebanon War with Israel and the subsequent civil war within the ranks of the PLO changed the PLO's institutions and factions in ways that concentrated more political power in both Arafat and the moderate elements of Fatah, the largest and most powerful faction within the PLO. Up until 1982, the PLO's decisions were made by consensus among a range of PLO factions (ranging from moderate to rejectionist) with autonomous institutional interests, power bases, and ideological affinities. No major concessions could be made that might jeopardize the political position of any faction (especially the rejectionist groups whose leaders had strong ties to radical Arab states). Moreover, Arafat's decisions were influenced by his desire to preserve Palestinian unity and maintain the PLO as the sole representative organization (Sahliyeh 1986: 92).[24]

The Israeli invasion and the forced evacuation of the PLO from Lebanon resulted in the loss by the PLO of its political and military infrastructure,

Table 5.3. An Israeli-Palestinian De-escalation Chronology

Date	*Event*
November 1974	Arab states at Rabat declare that the PLO is the sole representative of the Palestinian people.
September 1978	Egypt and Israel agree on Camp David Accords.
June 1982	Israel invades Lebanon.
August 1982	The PLO evacuates Beirut.
May 1983	The Fatah rebellion occurs in Lebanon.
December 1983	Arafat and 4,000 remaining PLO personnel evacuate from Lebanon.
February 1985	Jordan and the PLO agree to an accord on negotiations with Israel.
December 1987	The intifada (Palestinian uprising) begins.
July 1988	King Hussein announces Jordan's disengagement from the West Bank.
November 1988	The PLO declares an independent state of Palestine and formally recognizes Israel's right to exist while condemning terrorism.
December 1988	The United States opens a dialogue with the PLO.
March 1990	The National Unity government in Israel falls, followed by a new Likud right-wing government.
January 1991	The Persian Gulf War begins.
October 1991	The Middle East conference in Madrid takes place.
June 1992	Rabin and the Labor Party defeat the Likud in the Israeli national election.
January 1993	Secret negotiations begin in Oslo between the PLO and private Israelis.
September 1993	The PLO and Israel sign the Declaration of Principles in Washington, D.C.

Source: Based partially on Kriesberg's (1998: 191) chronology of events.

which no other Arab country would have permitted to be built. Prior to 1982, the PLO had enormous freedom in Lebanon, operating as a state within a state. But without Lebanon as a political and military base of operations, the PLO was unable to continue its long-held strategy of engaging the Israelis militarily from a contiguous Arab country. It was too unlikely that guerrilla tactics could be used from remote territories in North and South Yemen, Algeria, and Sudan. Therefore, the potential of translating the PLO's military force into tangible political and bargaining power was remote if nonexistent.

Meanwhile, Arafat and other moderates in the PLO believed that the Lebanon War demonstrated the futility of armed struggle. They advocated a more aggressive diplomatic posture, intensifying contacts with the West (particularly the United States) while conveying a readiness to recognize Israel's right to exist. Their aim was to facilitate efforts by the West to exert pressure on Israel to make meaningful concessions to the Palestinians (Sahliyeh 1986: 93). These policies polarized political opinion within Fatah and eventually led to full-scale war among Fatah organizations in Lebanon.[25]

Actual fighting broke out between Arafat supporters and opponents in June 1983 with the explicit approval of the Syrians, who wanted to eradicate altogether Arafat's military presence in Lebanon. As the fighting continued into the fall of 1983, Arafat's troops lost important strongholds in refugee camps in Syrian-controlled areas. Finally, in November Saudi Arabia pressured the Syrians to end the fighting and to allow Arafat and his troops to leave Tripoli peacefully. In December 1983, more than four thousand Arafat supporters were transferred out of Lebanon (Sahliyeh 1986: 160–175).

After Arafat's exile in 1983, three major internal changes occurred within the PLO that produced significant changes in the PLO's strategy. First, the radical commando groups were removed from Fatah, thereby increasing Arafat and Fatah's ability to pursue a diplomatic option while marginalizing Syria's hard-line influence in the process. Second, Arafat's power and influence were consolidated in 1984 at the seventeenth Palestinian National Council when moderate elements beat back proposals from radical groups for organizational reforms that would have limited Arafat's authority.[26]

Last but not least, the wishes and interests of the Palestinians in the West Bank and the Gaza Strip were prioritized over Palestinian interests in the diaspora. These "inside" Palestinians became Arafat's primary constituency. And, since the "inside" Palestinians endorsed territorial compromise with the Israelis, Arafat and the PLO began to adopt an even more flexible foreign policy. At this point, Arafat explicitly broke away from the principle of consensus building in favor of majority rule (Groth 1995; Hassassian 1997).

THE INTIFADA

The intifada was an internal uprising of Palestinians in the West Bank and the Gaza Strip against Israeli occupation. The campaign of civil disobedience—

demonstrations, strikes, tax revolts, and boycotts—was conducted by a broad sector of Palestinian society that included students, teachers, labor unionists, construction workers, shop owners, and farmers as well as the poor and the unemployed. The campaign started in December 1987 and was sustained by a highly mobilized public through 1989 (Hunter 1993).

The intifada imposed on Arafat and the PLO the urgent need for new, decisive actions that reflected the interests of local Palestinians in the West Bank and Gaza who demanded more progress toward negotiations with the Israelis. The PLO's leadership was aware that its lack of success could lead to a loss of influence over the territories; more than anything else it feared that Israel and the United States would strike a deal with the inside leaders at the expense of the outside Palestinians (Groth 1995; Muslih 1997).

In order to ensure their control over the territories, Arafat and the mainstream PLO leadership believed that a formal dialogue with the United States was necessary to bring about negotiations with the Israelis. The question was how to overcome U.S. opposition to the PLO's participation in the peace process. This question became more critical when Jordan announced its formal disengagement from the West Bank—opening up a real possibility that the Israelis might annex the territories unless the PLO took a strong stance quickly (Sayigh 1997: 622–624).

Therefore, at the meeting of the Palestinian National Council in November 1988, not only did the PLO members announce their formal declaration of independence, but they approved UN Resolutions 242 and 338 and condemned the use of terrorism. This policy committed the PLO to support three major measures: an acceptance of Palestinian borders within Gaza and the West Bank (as opposed to the original pre-1967 borders) as a Palestinian state; the recognition of Israel's right to exist; and a preference for diplomacy and a peaceful settlement with Israel (rather than an armed struggle) (Muslih 1997: 46–48). One month later and after a more explicit public statement by Arafat supporting these measures, the United States opened a formal dialogue with the PLO (Sayigh 1997: 624).[27]

A SHIFT IN SOVIET STRATEGIC BACKING

The shift in Soviet support from 1985 to 1993 deprived the PLO of a global ally. Although the Soviets never really had much influence over Fatah or Arafat's policies, they supported the PLO largely because of its competition with the United States in the area (Golan 1997: 124). By the end of the 1980s, however, Gorbachev had come to power advocating a new approach, the "de-ideologization of foreign policy," which led to a reduction of Soviet support for national liberation movements in the Third World, including armed struggle and terrorism. Gorbachev argued that it was time to end superpower competition in regional conflicts. In 1988, during Arafat's visit to Moscow, Gorbachev advised Arafat to accept Israel's right to recognition and security, along with the Palestinians' right to self-determination. Gorbachev sent the same message to several other visiting PLO delegations as well (Golan 1997: 126–127).

Nonetheless, the most serious development from the PLO's point of view was the warming of relations between Israel and the Soviet Union, in particular the influx of new Jewish immigrants from the Soviet Union. Between 1989 and 1992, 375,000 immigrants arrived in Israel from the former Soviet republics and increased Israel's total population by 7 percent (Fein 1995: 162). The PLO was especially alarmed because the influx of immigrants would lead to massive expropriation of Palestinian lands and a wave of new Jewish settlements in the territories. These developments could only help to consolidate Israel's hold over the territories (Sayigh 1997: 639). The emigration issue dominated PLO-Soviet relations from 1990 to 1993 while PLO officials tried to convince the Soviets that Jewish emigration threatened the peace process and reinforced the hard-line policies of the Israeli government (for example, Shamir's administration). As the growth of Jewish settlements increased dramatically during this period, leaders of the PLO felt enormous pressure to take decisive actions that would protect not only the "inside" Palestinians but their political legitimacy within the territories (Groth 1995: 25).

THE FINANCIAL CRISIS OF 1990–1993

The progress of secret talks between the PLO and Israel at Oslo in the summer of 1993 was accompanied by severe financial strain within the PLO. Due to the PLO's support for Saddam Hussein during the Persian Gulf War, the Gulf Cooperation Council states cut aid to the PLO formally in March 1991, but the cuts began as early as 1990.[28] The PLO's budget declined more than 50 percent, from $320 million in 1990 to $140 million in 1992. The flow of funds to the occupied territories also plummeted from $120 million in 1990 to $45 million in 1992 (Sayigh 1997: 656). The situation was aggravated further when an estimated three hundred thousand Palestinians in Kuwait were thrown out of work by the crisis, leading to the loss of approximately

$80 million per year in remittances to families in the West Bank (Rubin 1994: 157). Meanwhile, the Gulf crisis and wartime curfews in 1990–1991 also aggravated the economic conditions inside the territories. More than one-third of an annual GNP of $1.6 billion was lost. In the spring of 1993, Israeli border closures following bombings by Hamas caused further sharp drops in employment and reduced Palestinian income by $2 million daily (Sayigh 1997: 656).

The deteriorating economic conditions led to a disagreement between the PLO and "inside" Palestinians over the next course of action. Local Palestinians began to establish their own diplomatic connections with the United States and Israel, both of which were actively soliciting an alternative Palestinian leadership that would prove to be more flexible. Meanwhile, the PLO faced still more challenges from Hamas, which remained outside the Fatah framework and portrayed itself as an alternative representative for a large portion of the Palestinian people. By the summer of 1991, divisions between Fatah and Hamas were serious enough to provoke a series of clashes between their followers. Arafat feared that without any progress toward negotiations, Hamas would accelerate its influence in the territories at the expense of the PLO (Abu-Amir 1993: 5–19).

Reluctantly, Arafat allowed PLO officials to accompany an "inside" Palestinian delegation to the Madrid Conference in 1991 and subsequent bilateral negotiations with the Israelis in Washington, D.C. However, he obstructed and delayed any agreement not signed directly by the PLO and himself between 1991 and 1993. As dissension emerged between Arafat and the "inside" Palestinians over the terms of a peace settlement, Arafat moved aggressively to seal a deal with the Israelis in secret talks in Oslo (Sayigh 1997: 654–655).

Arafat preferred the talks in Oslo over the sessions in Washington, D.C. for two major reasons. First, the negotiations in Oslo were purely between pro-Arafat officials and the Israeli Labor Party, ignoring their respective opposition groups and the input from Arab states and the United States. Second, the negotiations in Oslo gave the PLO Israeli recognition as the sole legitimate representative of the Palestinian people (Groth 1995: 59).

The Oslo Accords were signed eventually by the PLO and Israel in a secret meeting on August 20, 1993.[29] Although the PLO was aware of the limitations of the Oslo Accords, it had little choice but to accept them. Its position in the post–Cold War and post–Gulf War era was very much weakened and thus it had consciously come to terms with Israel in order to secure a foothold on Palestinian soil before its regional and international standing eroded any further. The financial crisis, administrative and organizational breakdown, and political dissent during the previous two years underscored the precariousness of the PLO environment (Sayigh 1997: 660).

For Arafat, the Declaration of Principles (signed in September 1993) offered two significant strategic advantages: Israeli recognition of the PLO and the guarantee that the PLO would emerge as the sole governing authority in the territories. Arafat was willing to make short-term territorial concessions to Israel and abandon the Palestinian diaspora in order to ensure the PLO's political survival (Sayigh 1997: 660). In the post-Oslo period, the extension of the PLO's influence into the new Palestinian Authority occurred very quickly while local Palestinian leaders and outside Palestinians were either coopted or marginalized. The result was the consolidation of a neopatrimonial system with Arafat as the sole leader (Sayigh 1997: 663).

Arafat as a Policy Entrepreneur

The shocks in 1982 (Israeli invasion), 1983 (Fatah rebellion), 1987 (intifada), and 1990–1991 (changes in Soviet and Arab patronage) provided Arafat with opportunities to consolidate his personal power within the PLO while directing the PLO's policy toward a negotiated settlement in 1993. Before 1982, the major challenge facing Arafat was to maintain unity among disparate factions and to do so in the presence of constant intervention by one Arab state or another. This gave rise to the politics of consensus rather than majority rule, since the outvoted group could seek external support and threaten the PLO's claim to be the sole representative of all Palestinians (Sayigh 1997: 679).

Consensus politics granted disproportionate influence over decision making to the smallest group so long as it had a seat on PLO bodies and, by extension, to the Arab state backers of proxy groups such as the Syrian-sponsored Sa'iqa and the Iraqi-sponsored Arab Liberation Front. There was also very little incentive to deepen the PLO's unity by merging the different factions since each group could lay claim to a share of the PLO's funds and appointments according to an agreed quota (Sayigh 1997: 679).

The shocks of 1982 and 1983 diluted the influence of radical Palestinian leaders who lost their power bases in Lebanon. At that point, Arafat began to weaken his remaining colleagues and potential rivals in order to consolidate his own political position within the PLO. For instance, he concentrated formal authority for a growing number of the PLO's departments and programs in his own hands, merging the military and finance sections of Fatah and the PLO on the one hand, and duplicating agencies that had not yet come under his control on the other. He also fragmented organizational structures and channels, relying instead on increased distribution of patronage to maintain his personal control (Sayigh 1997: 688). By 1991, the PLO's institutions would include only those factions that did not participate in the rebellion against Arafat in 1983 (Groth 1995: 79).

During the intifada in April 1989, Arafat secured the nomination from the PLO Executive Committee and Central Council as president of the recently declared State of Palestine. Although the post had little real political power at the time, later it would enable Arafat to stand above the PLO and Fatah when necessary (as in the case of the Oslo negotiations) and marginalize both structures (Sayigh 1997: 633). Four months later in August 1989 Arafat also consolidated his political control over Fatah by insisting on two new posts—Fatah commander-in-chief and chairperson of Fatah's central committee—to which he was selected. He then acted to dilute the influence of the core founding members of Fatah by expanding the institution's central committee and revolutionary council with a large number of second echelon leaders and middle-ranking cadres dependent on his personal patronage (Sayigh 1997: 633–634).

Meanwhile, Fatah's leadership within the PLO was eroding as Arafat's political influence was increasing. The deaths of key Fatah leaders in 1982, 1988, and 1992 increased the balance of power in Arafat's favor, particularly when they were not replaced by other senior leaders but rather with committees of several individuals who were much easier to manipulate. The remaining core members of Fatah, who had no constituencies of any significance in the territories, were no match for Arafat (Groth 1995: 21).

The financial crisis in 1990–1991 weakened the PLO as an organization but strengthened Arafat's personal leadership. In order to meet the exigencies of the crisis, Arafat reduced or slashed budgets for welfare, PLO institutions, armed forces, and diplomatic stations abroad. These policies revoked the social contract between the PLO and its traditional constituencies in Lebanon, Jordan, Algeria, Yemen, and other Arab countries—constituencies that were least likely to approve Arafat's post-1991 policies. Hence, the financial crisis enabled Arafat to complete a process that he had begun in

1982, namely, a restructuring of the PLO's institutions to suit an "occupied territories first" policy (Groth 1995: 75–76).

Israeli Shocks

THE PERSIAN GULF WAR OF 1991

Rather than softening the hard-line strategies of Prime Minister Shamir, the war reinforced his views that the Palestinians were hostile to Israel and that peace was not possible (Aronoff 1998: 31). Despite this position, Shamir participated in bilateral negotiations with a Palestinian delegation from the territories only after he came under intense pressure from the United States. He feared that his refusal would result in both the loss of badly needed U.S. loan guarantees and good relations with the United States, which were already under intense strain (Stein 1993b: 23). Nonetheless, he had little intention of reaching a settlement with the Palestinians. Shamir even went so far as to say that he had planned to drag out the negotiations for at least ten years in order to increase the number of settlers in the territories and nullify any possibility of a separate Palestinian state (Segev 1998: 44).

While the Persian Gulf War had little effect on changing Shamir's expectations about de-escalation, it had a profound influence on Yitzhak Rabin, Israel's future prime minister. According to one of his close advisers, Rabin believed that the war demonstrated that the Israeli population had little desire to incur more deaths in future wars and that the increased willingness to emigrate in order to avoid such losses were clear indicators that a de-escalation strategy was required (Aronoff 1999: 32).

THE INTIFADA AND LIKUD'S POLICY MONOPOLY

The intifada did little to change the expectations and strategies of the Israeli government toward the PLO. From 1977 to 1992, the Likud Party had dominated Israel's foreign policy, including the two coalition governments with the Labor Party from 1984 to 1990. Historically and ideologically, Likud was committed to permanent control over the territories. Although leaders of Likud were willing to negotiate the terms of some form of Palestinian autonomy, they were unwilling to cede military control or give up their right to expand Jewish settlements in the area. Leaders of Likud were also adamant that if negotiations were to occur the discussions would be conducted exclusively with Palestinians from the "inside" territories, not with the PLO or any other delegations from the outside (Barzilai 1996; Telhami 1996).

The political impasse between the Israelis and the Palestinians would end eventually but not until the national election of 1992 when the Likud Party was defeated by a coalition of Labor and leftist/peace-oriented parties. The outcome in the national election of 1992 was not only unanticipated; it was considered a major political turnover (Elazar and Sandler 1995). The Likud Party had dominated the Israeli government since its own upset victory over the Labor Party in the national elections of 1977 and established a long policy monopoly over Israeli policy in the territories. Labor's return as the ruling party in the national elections of 1992 paved the way for the Olso Accords of 1993. In fact, analysts argue that it was the "sufficient" condition for rapprochement between Israel and the PLO (Telhami 1996; Aronoff and Aronoff 1998).

CHANGES IN ISRAELI PUBLIC OPINION AND THE END OF LIKUD'S MONOPOLY

There is little doubt that the intifada had a dovish impact on Israeli public opinion, a development that was carefully observed by the Labor Party but largely ignored by the Likud Party. Survey results (Goldberg et al. 1991; Arian et al. 1992; Arian 1996) showed that the intifada heightened the public's sense of threat and increased popular support for a strong, harsh military response to the uprising. In the long run, however, the uprising had a steady moderating (or dovish) influence on the public's views about the future of the territories (Goldberg et al. 1991: 5, 58; Arian 1996: 78). For instance, surveys showed that in 1984 only 35 percent of Israelis polled were willing to give up the territories in exchange for peace. By 1993, 54 percent of Israeli respondents were endorsing this position (Arian 1996: 99).

The gradual trend toward conciliation was also evident regarding the public's attitudes on the establishment of a Palestinian state. In 1987, a survey of Israeli citizens showed that 21 percent of the respondents supported a Palestinian state in comparison to 37 percent who favored it in 1994 (Arian 1996: 107). Meanwhile, negotiating with the PLO became a more acceptable option. In 1984, 21 percent of those surveyed supported negotiations, whereas 60 percent favored this option in 1994 (Arian 1996: 107).

Public opinion polls taken in early June 1992 (prior to the national election) showed that significant portions of the Israeli public indicated that the most important issue in choosing a party was its platform on the issues of security and the peace talks. Likud and Labor devoted their main efforts at capturing the votes in the center of the Israeli political spectrum. Unfortunately for Likud, voters evaluated Labor as the party that was most likely to end the intifada and achieve peace with the Arabs (Arian and Shamir 1995; Shachar and Shamir 1995).[30]

The Labor Party was able to capitalize on the changing trends in the Israeli electorate because it, too, had been moving in a moderate direction in the 1980s, particularly since the Israeli invasion of Lebanon. One obvious reason for this shift was Labor's attempt to highlight its differences with Likud (Inbar 1991). The dovish tilt in Labor's platform forced the hawkish elements of the party to leave, making room for dovish leaders to move into key leadership positions. Democratic reforms within the party also ensured the arrival of younger, more flexible leaders into important leadership roles (Aronoff and Aronoff 1998).

When in 1988 the PLO announced its formal recognition of Israel's right to exist and its renunciation of terrorism as a policy instrument, Labor leaders began to put greater pressure on Rabin (as defense minister) to make territorial concessions, recognize the Palestinian national movement, and be more open to a Palestinian state (Inbar 1991; Aronoff and Aronoff 1998). These policy positions correlated with the increasing trend of public support in 1992 for negotiations with the PLO, participation in an international peace conference, and territorial concessions that might lead to a Palestinian state (Arian 1996). Labor, therefore, was able to win over Likud in 1992 because it had carved out a policy position that was associated more closely with bringing about a peaceful agreement on the Palestinian problem that would ensure security and reestablish warmer relations with the United States.

Rabin as a Policy Entrepreneur

Yitzhak Rabin and the Labor Party's electoral victory in June 1992 marked the first time in fifteen years that Likud was completely excluded from power. The public perception of Rabin as a security-minded centrist won him the support of crucial "swing" voters in the middle of the Israeli political spectrum. Rabin ran on a political platform that stressed the theme that Palestinian terrorism did not constitute a threat to the country, but rather was an issue of personal security, which would be enhanced not by fighting the Palestinians but by striking a deal with them (Makovsky 1996: 85).

In September 1993, Rabin negotiated a final deal with the PLO that opposed the basic principles outlined by Menachem Begin, the Israeli prime minister, and the Likud Party in the Camp David Accords of 1979 (for example, territorial withdrawal and recognition of the PLO). However, from 1967 to 1987, Rabin had consistently opposed any negotiations with the PLO and the establishment of an independent Palestinian state. He had adhered to one particular solution: territorial compromise in the context of an Israeli-Jordanian confederation (Ben-Yehuda 1997: 207). What accounts for his change in thinking?

The intifada played a significant role in bringing Rabin to the realization that the status quo was untenable (Aronoff 1999; Bar-Simon-Tov 1999). He became convinced that the uprising was being conducted by a politically aware Palestinian population that was willing to carry out an uncompromising struggle over the long term. He also noticed a trend of increasing Palestinian radicalization and growing support for Islamic fundamentalist organizations. Persuaded that the intifada was a "continuation of the Israeli-Arab conflict by other means," Rabin argued that a political agreement had to be sought through compromise (Ben-Yehuda 1997: 210).

After his national election victory in 1992, Rabin initially believed that the PLO's political and financial weakness would make it more malleable and enable him to achieve the agreement he had believed possible in 1989—a deal with the Palestinians in the territories rather than with the PLO leadership in Tunis.[31] By early 1993, Rabin had concluded that the "inside" Palestinians could not and would not act independently of the Tunis-based PLO leadership. He then turned to direct talks with the PLO in Oslo. By this time, Rabin believed that the international and regional changes offered Israel a short-term "window of opportunity" to resolve the core conflicts between Israel and its Arab neighbors. He argued that it was time for Israel to join the international movement toward cooperation and conciliation in order to enhance Israel's long-term security (Makovsky 1996: 112).

Meanwhile, Rabin perceived a new threat emerging in the territories—Hamas. He feared that the balance of power was shifting in favor of the radicals over the PLO, whose organization was bankrupt, financially and morally, in comparison. Senior military and domestic intelligence officials repeatedly pointed out that the PLO's institutions were withering, especially in Gaza, while Hamas was stepping in to fill the void. Rabin decided to strike a deal with the PLO, which then would have the primary responsibility of containing the radicals. Any future Israeli withdrawals from the territories would depend on Arafat's ability to ensure Israel's security (Makovsky 1996: 112–113).

In the final analysis, Rabin reached the conclusion that there would not be a solution to the conflict, or an end to terrorism, or enhanced security without a long-term separation between Israel and a Palestinian entity side by side. Although he did not go so far as to endorse the concept of a Palestinian state explicitly, in private conversations after the accords were signed, Rabin appeared to have accepted such an outcome. Rabin's willingness to accept the risks associated with a Palestinian state that could prove to be destabilizing to both Israel and Jordan was the most significant change in his strategic thinking (Makovsky 1996: 124; Inbar 1999: 161).

Finally, third-party involvement played some role in the Oslo outcome but not so much as a consequence of direct pressure to negotiate. Growing cooperation between the United States and the Palestinians threatened possible reduced cooperation between Israel and the United States should the Israelis decline to negotiate. The increased Israeli flexibility toward the Palestinians thus owes something to an indirect third-party pressure (Steinberg 1998; Segev 1998).

Multiple shocks (Table 5.4), then, combined with some expectational revision, third-party pressure, policy entrepreneurship, and reciprocity to yield substantial changes in Palestinian and Israeli willingness to negotiate. In this case, shocks not only contributed to changing expectations and strategies of key decision makers, they also created opportunities for policy entrepreneurs to change their institutional structures in ways that enabled the pursuit of de-escalatory policies. Arafat consolidated his political leadership within the PLO after each successive crisis while Rabin and the Labor Party capitalized on the dovish shift in Israeli public opinion by defeating the Likud Party in the national election of 1992. On the Israeli side, the joint influence of shocks and Rabin's influence on offering Israeli concessions were probably more significant than either the combined effects of shocks, third-party pressure, or reciprocity. In this case, the interaction between shocks and policy entrepreneurs is especially noteworthy.

Table 5.4. Shocks in the Israeli-Palestinian Rivalry, 1991–1993

Shift in external threat	*Shift in competitive ability*	*Shift in regime orientations/ strategies*	*Economic resource crisis*
Hamas (for both Israel and the Palestinians)	Israeli invasion of Lebanon, intifada, Gulf War, loss of Soviet patronage, infighting (Palestinians); intifada, Gulf War (Israel)	Defeat of Likud (Israel)	Financial crisis (Palestinians)

Table 5.5. Modeling the Outcome for the Israeli-Palestinian Rivalry, 1990–1993

Variable	*Israeli-Palestinian rivalry*
Shocks	Present
Expectancy revision	Present
Consolidated and new policy entrepreneur	Present
Third-party pressure	Present
Reciprocity	Initially
Reinforcement	Absent
Outcome	No change

Political shocks tell us where the important turning points are likely to be. They do not guarantee turning points. Rather, they delineate the moments at which decision makers are either most likely to change their minds or to gain office. Without the environmental challenge that shocks pose, inertia overrides the inclination of decision makers to pursue new and risky policies. But de-escalation is neither the result of immutable, large-scale forces or the actions of a few brave individual policymakers. A number of factors (Table 5.5) need to converge for changes to occur in protracted conflicts.

Nonetheless, de-escalation is never a single event. It is part of an ongoing process that must be maintained through continuing reciprocity and reinforcement. If it is not maintained, then the process is likely to revert to earlier modes and relationships. Divisions within the Israeli political community between right and left, paralleled within the Palestinian political community by divisions between moderates (PLO) and radicals (Hamas), intervened to make reinforcement less likely and quite short-lived in this case. The process could easily have broken down on its own, but the assassination of Rabin and the return to power of Likud ensured it. Twenty years later, Jewish settlements in Gaza have been abandoned, but they continue to expand in the West Bank. On the opposing side, the PLO and Hamas continue their feuding as well, with Hamas in control of Gaza.

CHAPTER 6

The Indo-Pakistani Rivalry, 1947–2010

The Indo-Pakistani rivalry has proven to be one of the most enduring conflicts of the post–World War II era. It has resulted in four wars (1947–1948, 1965, 1971, and 1999) and multiple crises, several of which almost culminated in war. Since 1998, this rivalry has also acquired an overt nuclear dimension. Unilateral, bilateral, and multilateral efforts to bring it to a close have proven to be mostly fruitless. The origins of this rivalry have been discussed at length elsewhere (Ganguly 1994; Paul 2005). Suffice it to say that the conflict can be traced to the process of British colonial disengagement from the subcontinent in 1947 and the competing visions of state construction in South Asia. The two successor states of the British Indian Empire in South Asia, India and Pakistan, were created on the basis of divergent conceptions of nationalism. Indian nationalism was fundamentally secular and civic and Pakistani nationalism was ethnic. The two divergent visions clashed over the control of the Muslim-majority state of Jammu and Kashmir (Brines 1968).

This chapter will show how despite external shocks, third-party pressures, and the presence of change-seeking entrepreneurs there has been only one important time span (1972 to 1979) when the rivalry saw significant de-escalation. However, even during this period the underlying sources of discord remained unabated. Intransigence and a lack of reinforcement from the adversary as well as other exogenous shocks brought this period of de-escalation to a close. To that end, this chapter will explore efforts at de-escalation in 1962–1963, 1966, 1972, 1980, 1989, 1999, 2004–2007, and 2009–2010.[1]

Focusing on the evolution of the rivalry, the several attempts at its de-escalation are framed within a series of several distinct historical phases. The phases are 1947–1966, 1967–1971, 1972–1989, 1990–1998, and 1998 to the present. This periodization is not idiosyncratic. Compelling historical and structural factors can be advanced to justify it. Beginning in 1947, the initial phase of the rivalry culminated in the Indo-Pakistani war over Kashmir in 1965 and the subsequent Soviet mediation of the conflict in 1966. The next phase lasted from 1967 to the third Indo-Pakistani war of 1971 and the bilateral peace agreement of 1972. The third phase, a period of relative quiet, lasted until 1989. It metamorphosed after the Soviet invasion of Afghanistan in 1979 and lasted until the Soviet withdrawal in 1990. A new phase began in 1990 and lasted until the Indo-Pakistani nuclear tests of May 1998. It entered its most recent phase in 1999 with the Kargil war and continues until today. These periods have seen the rivalry wax and wane but have experienced only one viable move toward rapprochement. This move toward de-escalation came in the wake of Pakistan's overwhelming military defeat in 1971 and the Shimla Accord of 1972.

Phase One: The Origins and Evolution of the Rivalry

A brief discussion of the origins of the rivalry is necessary. The rivalry stems from competing visions of state construction in South Asia. The Indian nationalist movement was predominantly secular and had sought to create a civic, democratic polity. The Pakistani nationalist movement, in contrast, had sought to create a homeland for the Muslims of South Asia. The Muslim League, which had spearheaded the demand for Pakistan, had argued that despite the professed commitment to secularism, Muslims, as the largest minority community in the British Indian Empire, would be at a disadvantage in a predominantly Hindu India. In effect, they argued that the Indian National Congress, the dominant party of the Indian nationalist movement, had failed to make a credible commitment to protect the rights of Muslims in an independent India.[2]

These two competing conceptions of state construction clashed over the question of the division of the British Indian Empire following colonial disengagement. At the time of colonial withdrawal, two classes of states had existed in the empire. They were the states of British India directly ruled from Whitehall and the so-called princely states, which were nominally independent but had recognized the "paramountcy" of the British crown. With the departure of the British from the region, Lord Mountbatten, the last viceroy, decreed that the doctrine of paramountcy would lapse and that the princely rulers would have to choose between India and Pakistan. He also added that predominantly Muslim states, which were geographically contiguous, would join Pakistan.

The principal problem between India and Pakistan arose over the status of the state of Jammu and Kashmir. This state was predominantly Muslim, had a Hindu monarch, and shared borders with both India and Pakistan.[3] India laid claim to Kashmir to demonstrate that a predominantly Muslim state could exist within the confines of a secular polity. Pakistan (Bhutto 1969), with equal force, argued that without Kashmir, its identity as the homeland for the Muslims of South Asia would remain incomplete. These pristine commitments have lost much of their force over the past fifty odd years. Pakistan could not endure on the basis of religion alone. The rise of Bengali linguistic nationalism contributed to Pakistan's collapse in 1971. If religion alone could not provide the basis of state construction, then the Pakistani irredentist claim to Kashmir was basically meaningless. Similarly, India, though formally a secular state, has failed to guarantee the rights of its Muslim and other religious minorities on multiple occasions (Ganguly 2003). Consequently, to claim that India must necessarily hold on to Kashmir to ensure its secular status is equally flawed.

At the time of independence and partition, these principles, however, had much greater significance to elites of both sides. Thus, contrary to popular belief, Pakistan did not initiate the war over Kashmir in 1947–1948 for strategic or material reasons. It did so primarily because its founding elite believed in the importance of Muslim confraternity in South Asia (Khan 1970). Kashmir, as a predominantly Muslim state with a border contiguous to Pakistan, had to be incorporated into Pakistan.

The conflict ended with India referring the case to the United Nations Security Council under the terms of a breach of international peace and security. On January 1, 1949, the United Nations declared a cease-fire and both sides accepted it. Subsequently, the dispute became a subject of vigorous (and highly partisan) multilateral diplomacy at the United Nations until about 1960. Neither side, however, showed much willingness to cede ground and the issue quickly became deadlocked.

During this period, three critical shocks, two exogenous, the other endogenous to the region, led to the intensification of the rivalry.[4] The first came as early as 1954 with the Eisenhower administration's decision to make Pakistan a key ally in the strategy of containment.[5] The interests of the two sides, of course, were far from convergent. Pakistan sought to bring the United States into the region to balance Indian power. The United States, however, convinced itself that Pakistan, as a predominantly Muslim nation, would serve as a bulwark against Communist expansionism.[6] President Eisenhower's oft-cited letter in which he sought to allay Prime Minister Nehru's concerns about U.S. weaponry being used against India offered little solace to either Nehru or his cabinet (Gould and Ganguly 1992). Even though India's defense spending did not dramatically increase in response to the growth of Pakistani military power, the tenor of the relationship certainly worsened.[7]

The second shock came a mere four years later. As Pakistan's first constitutional order proved far too brittle the military seized power in 1958. The precise social and political forces that contributed to the seizure of power are not a matter of immediate concern. They have been discussed at length elsewhere (see McGrath 1996). The end of Pakistan's brief experiment with democracy deeply disturbed Nehru, still the principal architect of India's foreign policy and a committed democrat (Brown 2003). Nehru correctly feared that a military regime would prove to be even more intransigent in its dealings with India. Misgivings about Pakistan, which were already rife amid India's foreign policy elite, simply deepened.

The third shock that profoundly deepened the rivalry was the Chinese attack on India's northern borders in 1962.[8] The impact of this exogenous shock to India's foreign and security policies cannot be overstated. Even though India's leaders continued to rely on the hoary language of Third World solidarity and nonalignment, the ideational elements of India's policies started to wither away. Indian decision makers increasingly came to the realization that moral suasion counted for little in the harsh world of global politics. Nehru's vision of an alternative global order based upon an abiding faith in multilateral institutions, the creation of new global norms, and self-imposed restraints on the use of force was all but abandoned.

The Shock of 1962 and Thereafter

In October 1962, the Indian Army suffered its most dramatic military defeat at the hands of the Chinese People's Liberation Army. Troops who were ill-equipped, inadequately armed, and unprepared for high-altitude combat faced a battle-hardened and carefully deployed Chinese army with disastrous consequences along India's Himalayan border.[9] The attack came as a dramatic surprise to the Indian political-military establishment and resulted in a near complete rout of the Indian armed forces (Palit 1991). In the aftermath of this conflict, a serious debate ensued in India about the wisdom of low defense expenditures and the pursuit of a policy of nonalignment. Despite a vigorous debate, the country chose not to abandon its nonaligned policies but did seek limited military assistance from the United States.[10] Also, it embarked on a significant program of defense modernization. India decided to acquire an air force of forty-five squadrons equipped with supersonic aircraft, it decided to raise the manpower ceiling of the Indian army to a million men, and it chose to raise ten new mountain divisions trained and equipped for high-altitude warfare. Finally, it set in motion a limited program of naval modernization.

India was briefly militarily dependent on the Western powers, most notably the United States and the United Kingdom, for military assistance. Owing to the Cold War exigencies of courting Pakistan, the United States and the United Kingdom exerted pressure on India to settle the Indo-Pakistani dispute on terms favorable to Pakistan. As a consequence of this pressure, the two sides did hold a series of negotiations in late 1962 and early 1963. These discussions almost culminated in a settlement of the Kashmir dispute on the basis of a territorial division. However, owing to the intransigence of the principal Pakistani negotiator, Zulfikar Ali Bhutto, the talks ended in a deadlock (Kumar 2005). The specific reasons underlying Bhutto's unwillingness to compromise remain unclear. However, it may have been rooted in his belief that India was in a fragile state in the wake of the border war with China in 1962 and that subsequent military pressure might induce India to make territorial concessions in Kashmir (Bhutto 1969).

In the absence of the external shock of the war with China in 1962 and India's acute dependence on the Western powers for military assistance, it is most unlikely that Prime Minister Jawaharlal Nehru would have consented to holding talks with Pakistan. Yet despite the talks that were held, deescalation was undermined because of the intransigence of key individuals and the lack of reciprocity. The Indian political leadership then (and even today) would readily settle for a territorial division of the state that reflected the disposition of forces at the end of the conflict of 1947–1948.

The very insubstantial U.S. arms assistance coupled with India's ambitious military modernization plans caused alarm in Islamabad. Pakistani decision makers promptly suggested that these U.S.-supplied weapons, ostensibly directed against a Chinese threat, would be used against Pakistan. This argument was both disingenuous and self-serving. India was now a status quo power with respect to territory and faced a palpable threat from

China. However, India's military modernization would indubitably undermine any Pakistani military efforts to wrest Kashmir back through the use of force.[11] Not surprisingly, the Pakistani military dictatorship of Field Marshal Mohammed Ayub Khan devised an elaborate plan to try and seize Kashmir through the use of force. This plan, Operation Gibraltar, culminated in the second Indo-Pakistani war of 1965. The war, which was brief, ended mostly in a military stalemate with India making some small territorial gains in Kashmir. These gains, however, were ceded to Pakistan in the Tashkent Agreement of 1966, which restored the status quo ante (more details can be found in Ganguly 1990).

Negotiations at Tashkent

The bilateral negotiations at Tashkent transpired because of third-party intervention. Specifically, the United States chose not to broker a postwar settlement, thereby enabling the Soviets to step into the breach. Accordingly, Leonid Brezhnev invited Prime Minister Shastri of India (who had succeeded Nehru in 1964) and President Mohammed Ayub Khan of Pakistan to the then Soviet Central Asian city of Tashkent for a postwar negotiation. Thanks to Soviet cajolery the two sides managed to reach an agreement. Under the terms of the Tashkent Agreement of 1966 India chose to return the strategic Haji Pir Pass over the objections of the Indian army, restoring in effect, the status quo ante. Furthermore, both parties also agreed to eschew the use of force in settling the Kashmir dispute.

Once again the exogenous shock of the war and the eagerness of the Soviets to enter the subcontinent provided the basis for this attempt at de-escalation of the conflict. Though Soviet intervention provided the two parties with a venue to resolve their immediate differences it did not address the underlying causes of the dispute or either side's expectations. Consequently, the effects of this round of de-escalation were mostly cosmetic excepting the return of the Haji Pir Pass.

Phase Two: The Rivalry Worsens

Despite the military stalemate that resulted from the 1965 war, Pakistani political-military elites remained unreconciled to the status quo in Kashmir.

With the imposition of a U.S. arms embargo on Pakistan following the war, however, they were no longer in a position to embark on yet another conflict with India.

The reduction in bilateral tensions by default proved to be quite short-lived because internal unrest within East Pakistan spilled over into India in 1971, culminating in the third Indo-Pakistani conflict.[12] The third war did not start over the question of Kashmir. Instead it arose from the dynamics of Pakistan's internal political processes. The crisis stemmed from the structural imbalances that had long characterized the two wings of the Pakistani state. East Pakistan, for all practical purposes, had been treated as an internal colony of West Pakistan. The consequent grievances of the East Pakistanis ultimately broke through in the first free elections that Pakistan held in 1970. A regional, pro-autonomy party, the Awami League, won an overwhelming victory in East Pakistan. The Pakistani military establishment, as well as the Pakistan People's Party, which was based in West Pakistan, were utterly unprepared for this outcome. As protracted and convoluted negotiations over power sharing collapsed in March 1971, the Pakistani army resorted to a brutal and callous campaign of mass murder in East Pakistan, designed to cow the attentive public. Faced with this reign of repression, over the next several months some ten million individuals took refuge in India.

By May 1971, Indian authorities had decided that India was not in a position to absorb the refugees into its already large population. They also realized that barring some dramatic action on the part of the global community, Pakistan would not create conditions conducive for the return of the refugees (Jacob 1997). Accordingly, Indian decision makers swiftly forged a political-military strategy to alert the global community of the crisis. Simultaneously, it carefully laid plans for an invasion of East Pakistan in the event diplomatic initiatives did not produce the desired results. After some months of feverish diplomatic activity that yielded little or nothing concrete, Indian decision makers decided that they would invade East Pakistan with the goal of establishing a separate state. In preparation for this endeavor they signed a treaty of "peace, friendship and cooperation" with the Soviet Union. The treaty, among other matters, committed the two parties to come to each other's assistance in the event of an attack by a third party. The treaty effectively neutralized any possible military threat from the People's Republic of China.[13] With the diplomatic initiatives exhausted, the treaty in place, and its military forces appropriately arrayed, India embarked upon a short, swift war that resulted in the rout of the Pakistani forces in East Pakistan within a span of two weeks. The concomitant breakup of Pakistan contributed to the creation of the state of Bangladesh.

In the aftermath of the conflict, India and Pakistan negotiated a bilateral agreement at the Indian hill station of Shimla in 1972 that normalized diplomatic relations, contributed to the return of ninety thousand Pakistani prisoners of war, and reiterated their mutual commitment to the non-use of force to resolve the Kashmir dispute. It also changed the nomenclature of the U.N.-sponsored cease-fire line to the Line of Control, reflecting a shift to bilateralism in Indo-Pakistani relations.

The inferences that the Pakistani military made from this dramatic military defeat are worthy of discussion. Instead of recognizing that their flawed policies had led to the military debacle they chose instead to blame the key civilian decision makers, most notably Zulfikar Ali Bhutto. Bhutto's culpability in precipitating the crisis of 1971, of course, is indisputable (Zaheer 1994). That, however, could not exonerate the military for the long years of gross and abject neglect of East Pakistan, of the systematic discrimination against Bengalis in the armed forces and the civil service, and the reign of repression they had unleashed in East Pakistan in March 1971.

Nonetheless, India had emerged as the clear, undisputed winner of this war and also as the dominant power in the region. Pakistani decision makers, while still unprepared to accept the territorial status quo in the region, nevertheless were forced to come to terms with the dramatic asymmetry in mutual power relations. Under the leadership of Prime Minister Zulfikar Ali Bhutto Pakistani decision makers decided to embark on a covert nuclear weapons program. In their view (Ahmed 1999), the acquisition of a nuclear weapons option would compensate for Pakistan's fundamental conventional asymmetry.

This time an interaction between endogenous and exogenous shocks resulted in a significant de-escalation of the rivalry. The endogenous shock emerged from the uprising within East Pakistan, and the military rout at the hands of its principal adversary, India, constituted the exogenous shock. Pakistan suffered a major military defeat and, in effect, emerged as the weaker power in South Asia. Consequently, it was forced to reassess its foreign policy behavior. It could ill-afford to initiate conflicts with a neighbor that now possessed overwhelming conventional superiority and a much larger economy. Consequently, Indo-Pakistani relations assumed a semblance of normalcy between 1972 and 1979.

Phase Three: A Stable Peace in South Asia?

In the aftermath of the Shimla Accord of 1972, the manifest features of the rivalry largely abated, even though the rivalry's underlying sources remained unaddressed. When the Soviet invasion of Afghanistan took place in 1979, an important opportunity for Indo-Pakistani rapprochement opened up. The two sides could have, through adroit diplomacy, managed to present a common front against the Soviet intrusion into the region's affairs. Indeed, to this end, Prime Minister Rajiv Gandhi dispatched Narasimha Rao, his minister for external affairs and a veteran Congress politician, to reassure General Zia-ul-Haq that India would not seek to exploit Pakistan's emergent security dilemma. His effort, however, was both unheralded and fruitless, as General Zia sought to exploit the adverse situation to bolster the tenuous legitimacy of his military regime. Instead of seeking a regional solution to the crisis he actively courted the United States largely in an effort to obtain economic and military largess.[14] More to the point, he, as well as significant segments of the Pakistani political-military establishment, remained utterly unreconciled to the territorial status quo in South Asia. Additionally, the military deeply resented India's role in the creation of Bangladesh (Arif 1995). Many in the military establishment sought opportunities to exploit cleavages within India for its role in the breakup of Pakistan in 1971.

The Soviet invasion of Afghanistan in December 1979 significantly altered the security situation in South Asia (Bhargava 1983). Pakistan had been at odds with the Carter administration because of significant differences over human rights and nonproliferation issues. However, with the Soviet presence in Afghanistan the administration in its final year did an abrupt about-face. It anointed Pakistan as a "front line" state and offered $400 million in economic and military assistance.

Sensing that the Pakistani military dictator, General Zia-ul-Haq, would seek to exploit the Soviet presence in Afghanistan, Prime Minister Indira Gandhi sent her foreign minister, Sardar Swaran Singh, to Islamabad to proffer a regional solution to the crisis. In Islamabad Singh sought to reassure

India's principal adversary that it would not exploit Pakistan's emergent security concerns. These efforts, however, proved to be of little consequence. General Zia-ul-Haq showed scant interest in Prime Minister Gandhi's proposals and also dismissed President Carter's offer of assistance as "peanuts" (Duncan 1989).

Shortly thereafter, in 1981, with the advent of the Reagan administration, Pakistan became the recipient of considerable U.S. largesse. Not surprisingly, India turned to the Soviet Union for weaponry to counter Pakistan's military acquisitions. Any hope of de-escalation in tensions simply evaporated for the next decade. The Soviet invasion, a significant exogenous shock to the region, had held open the possibility of a de-escalation of tensions, especially in light of Indira Gandhi's efforts at reassurance. Yet the Pakistani military establishment under General Zia-ul-Haq displayed no interest in reciprocity.[15]

Even with substantial U.S. military assistance, Pakistan was unable to embark on a full-scale conventional assault on India because of significant and growing military asymmetries.[16] Nevertheless, Pakistan chose instead to exploit existing tensions within India. Accordingly, in the 1980s, for example, the military regime of General Zia-ul-Haq became deeply embroiled in training, supporting, and providing sanctuaries for Sikh insurgents in the Punjab. The roots of the insurgency were indigenous, but Pakistani assistance prolonged its duration and increased its intensity (Wallace 1995). Pakistani support for the Sikh insurgents coupled with India's growing frustration with its inability to suppress the insurgency contributed to a significant crisis in Indo-Pakistani relations in 1987. The Indian military exercise, Brasstacks, which was designed partly to signal India's ability to strike Pakistan despite the ongoing insurgency in the Punjab, helped trigger this crisis (Bajpai et al. 1995). Some tentative evidence (Ganguly and Hagerty 2005) suggests that India was deterred from escalating the conflict either horizontally or vertically for possible fears of a Pakistani resort to nuclear weapons.

Phase Four: Kashmir Redux

Pakistani decision makers, whether civilian or military, had refused to accept the political arrangement in Kashmir despite a professed commitment to the Shimla Accord of 1972.[17] The existing military balance, however, had simply rendered it impossible for Pakistan to seek to wrest Kashmir back from India through the use of force. It is entirely possible that the issue of Kashmir might have abated somewhat if an insurgency had not erupted in the state in December 1989. But that shock was preceded by another one: the death of General Zia-ul-Haq.

General Zia perished in a plane crash in July 1988. For various reasons, including U.S. diplomatic pressure, the Pakistani military chose not to cling to political power. As a consequence, Zulfikar Ali Bhutto's daughter, Benazir Bhutto, assumed political office after a landslide electoral victory.

In India, Prime Minister Rajiv Gandhi had succeeded his mother following her assassination in 1984. With the emergence of a democratically elected leader in Pakistan, he and others believed that an opportunity to reduce tensions had opened (Dixit 2002). Accordingly, he initiated talks with Prime Minister Benazir Bhutto of Pakistan. These talks initially generated three substantive agreements on nuclear confidence-building measures, international civil aviation, and a series of further meetings on matters related to border management, trade, and tourism (Dixit 2002: 268). However, they quickly fell victim to the exigencies of India's domestic politics.

In December 1989, an ethno-religious insurgency erupted in the Indian-controlled portion of the disputed state of Jammu and Kashmir. The origins of this uprising were indigenous and stemmed from the exigencies of Indian domestic politics (Ganguly 1997), in particular the growth of political mobilization in the state against a backdrop of steady political deinstitutionalization. The conjunction of these two forces proved to be a lethal amalgam. As a new, politically conscious generation of Kashmiris saw that institutional channels of dissent were blocked they resorted to violence to express growing political discontent. Once the insurgency had erupted, Pakistani decision makers, most notably in the military establishment, sought to mold, shape, and direct the course of the rebellion.

What was originally a spontaneous rebellion soon became an externally orchestrated extortion racket (Davis 1995). The Soviet withdrawal from Afghanistan and the subsequent Soviet collapse also aided Pakistan in its ability to support the insurgents.[18] Thousands of mujahideen whom the Pakistani military establishment had organized, trained, and supported, were now available for the conduct of yet another jihad.[19]

With this abrupt rise in anti-Indian sentiment across much of the Kashmir Valley, Benazir Bhutto came under considerable pressure from the Pakistani military as well as other hawkish forces within the polity to adopt a hostile posture toward India. Within weeks she made a series of intransigent public remarks that swiftly brought various ongoing discussions to a close. The possibility of de-escalation that had arisen was, yet again, lost.

Endogenous shocks were responsible for opening as well as closing this particular prospect of de-escalation. The first shock stemmed from the death of General Zia-ul-Haq and the concomitant emergence of a democratically elected regime in Pakistan. It was also helped by Rajiv Gandhi's assumption to the office of the prime minister upon the assassination of his mother. Neither Benazir Bhutto or Rajiv Gandhi had been a witness to the partition of the subcontinent and both were therefore more open to the possibilities of Indo-Pakistani accord.

They may not have been ardent change-seeking entrepreneurs, but they were nevertheless open to the possibilities of discussion and negotiation. However, Bhutto was especially vulnerable to structural and social forces with the Pakistani polity. Religious zealots, whom President Zia-ul-Haq had cultivated, and the military establishment were both arrayed against her. Consequently, once the sudden rebellion broke out in Kashmir, another endogenous shock, her room for political maneuvering was constrained and she felt compelled to publicly outbid the other strident voices from within Pakistan. Under these circumstances the de-escalation process, which was still in an incipient stage, came to a quick close.

Phase Five: From 1998 Onward

Another serious effort at the de-escalation of tensions in the subcontinent did not ensue for almost a decade. Ironically, it was the Indian and Pakistani nuclear tests of May 1998 that triggered the de-escalation process. After years of nuclear ambiguity, both India and Pakistan conducted nuclear tests in May 1998. The origins of their respective nuclear programs and their motivations have been discussed at length elsewhere (Ganguly 1999; Ahmad 1999). In the aftermath of the tests both states faced a spate of multilateral sanctions. More to the point, a number of important world leaders expressed grave concerns about the dangers of renewed conflict in the region culminating in nuclear war (Talbott 2006).

Probably with a view toward dispelling some of these concerns Prime Minister Atal Behari Vajpayee of India initiated a bus service linking the cities of Amritsar (in India) and Lahore (in Pakistan) in February 1999. He also chose to visit Pakistan as he inaugurated the bus service. The visit was significant for its symbolic and substantive elements. The symbolism of the visit was important because Vajpayee, the leader of a right-wing, Hindu chauvinist political party, had chosen to visit Pakistan. More to the point, he specifically chose to go to the Minar-e-Pakistan, the site where the Muslim League had passed the original resolution for the creation of Pakistan in 1940 (Chari, Cheema, and Cohen 2007: 121). At that venue, Vajpayee publicly affirmed India's commitment to the territorial integrity of Pakistan.

Beyond these symbolic gestures of note the Lahore Declaration also sought to promote confidence-building measures designed to prevent the accidental or unauthorized use of nuclear weapons and called for a resumption of a composite dialogue to address a range of outstanding differences. Despite these symbolic gestures and substantive commitments the Lahore peace process quickly unraveled.

This attempt at de-escalation came to a close because the Pakistani military chose to launch a military operation across the Line of Control (the de facto international border) in the Kargil region of the state of Jammu and Kashmir.[20] When the Indian military discovered these incursions in early May 1999 it responded with considerable vigor while keeping the ambit of the conflict carefully limited to the Kargil region. Unlike in past conflicts where India had resorted to horizontal escalation through the opening of a second front, on this occasion it chose not to do so. In all likelihood, India's unwillingness to expand the scope of the conflict stemmed from its knowledge that such a widening could contribute to an escalatory spiral resulting in a Pakistani threat to resort to the use of nuclear weapons (Ganguly and Kapur 2010). Through the steady, unwavering but calibrated application of force India prevailed in the war. Around mid-July India had succeeded in ousting all the Pakistani intruders.

Significant external pressures that the nuclear tests had generated were the principal reasons for the genesis of this attempt at de-escalation. In effect, the overt nuclearization of the region had amounted to a radical change in its security milieu. Ironically, then, the tests need to be seen as the shock that prompted the pressures for de-escalation. Unfortunately, the fragmented power structure within Pakistan where the military had remained *primus inter pares* despite the presence of a civilian regime contributed to the lack of reciprocity and the failure of de-escalation. Worse still, it had actually contributed to an escalation of tensions through the Kargil misadventure.

The nuclearization of the region did not inhibit the Pakistani political-military establishment from making a renewed attempt to revive the Kashmir issue. On the contrary, it was emboldened in its resolve. It had correctly concluded that the fear of escalation to the nuclear level would inhibit the Indian security forces from expanding the scope of the conflict either vertically or horizontally (Ashraf 2004). In April 1999 Pakistani forces disguised as local tribesmen made a series of bold forays across the Line of Control in the Kargil district of Kashmir at altitudes of about sixteen thousand feet. Though they achieved tactical surprise, they failed to consolidate their gains. The areas where they made these incursions were extremely inhospitable, mostly snowbound for the bulk of the year, and were not easy to resupply. Once the Indian military discovered these incursions it launched a vigorous and unyielding campaign to dislodge the intruders. Within three months of the initial incursions the Pakistani forces had been pushed back to their original positions.

What was the principal purpose of these incursions? As argued earlier, by the late 1990s, India had succeeded in restoring a modicum of order to Kashmir. More to the point, the global community was increasingly losing interest in the Kashmir question. In large part, Pakistan chose to make these forays across the Line of Control to try and revive international interest in the Kashmir question. Mediation by the United States provided Prime Minister Nawaz Sharif of Pakistan a face-saving pathway out of this crisis (Talbott 2004). Despite the Pakistani withdrawal from the occupied positions, support for the Kashmiri insurgents did not come to a close. If anything, following a military coup against Prime Minister Sharif in October 1999, support for the insurgents increased.

Even after the tragic events of September 11, 2001, and U.S. pressure on Pakistan to end its ties to the Taliban regime in Afghanistan in particular and terrorist organizations in general, Pakistan refused to terminate its support for a range of insurgent groups operating in Kashmir. On December 13, members of two of these groups, the Lashkar-e-Taiba and the Jaish-e-Mohammed, launched a daring attack on the Indian parliament in New Delhi. Even though they managed to penetrate the security cordon of the parliamentary enclosure they were subsequently stopped and killed in the ensuing gun battle. In the aftermath of this episode, India embarked upon a large-scale military mobilization effort designed to coerce Pakistan to put an end to support for the insurgents. This exercise in coercive diplomacy, Operation Parakram, which lasted several months, nearly brought the two sides to the brink of war. Substantial U.S. engagement on both sides of the border played an important role in eventually defusing the crisis in the late summer of 2002 (Ganguly and Kraig 2005).

Continuing Deadlock: The Composite Dialogue

In 2003, Prime Minister Atal Behari Vajpayee again offered a resumption of talks with Pakistan. The motivations underlying his offer to hold talks with Pakistan remain unclear. It is possible that given the success of India's counterinsurgency strategy in Kashmir, which had managed to restore a semblance of political order and stability, he saw a renewal of talks with Pakistan as a possible pathway toward addressing the underlying sources of conflict. In any case, Pakistan reciprocated later in the year with the banning of two terrorist organizations and the imposition of a cease-fire along the Line of Control (Chari, Cheema, and Cohen 2007: 208).

By 2004, a full-scale, "composite dialogue" was under way with a range of issues extending from nuclear confidence-building measures to the demilitarization of the Siachen Glacier under discussion. According to one of the few accounts of the dialogue, considerable progress was made between 2004 and 2007 in addressing these issues. Unfortunately, domestic opposition to General Musharraf's regime, unrelated to his pursuit of this peace process with India, led to its petering out.[21]

With the civilian interregnum in Pakistan and a change in the key political actors the dialogue lost its momentum. Worse still, in November 2008, a group of ten terrorists affiliated with the Lashkar-e-Taiba attacked a series of targets in a well-coordinated fashion across Bombay (Mumbai), India's principal commercial and entertainment hub. For a host of complex reasons it took Indian security forces over three days to completely quell the terrorist attack (Rabasa et al. 2009). Despite the viciousness of the attack and India's ability to trace the origins of the terrorists to Pakistani soil based upon electronic intercepts, the Congress-led regime of Prime Minister Manmohan Singh avoided any resort to military action. In all likelihood India decided not strike Pakistan for two reasons. First, in the time that it took the security forces to bring the terrorists to heel any element of surprise was lost. It was entirely reasonable to surmise that Pakistani forces were in a suitable state of alert to repulse or respond to an Indian military strike. Second, the fear of nuclear escalation that had long dogged Indian military planning no doubt played a role in curbing an incentive to carry out military action. However, the regime in New Delhi felt compelled to end the ongoing dialogue.

What had initiated the "composite dialogue" in the first place? Vajpayee and Musharraf must be seen as change-seeking entrepreneurs. Though Vajpayee made the initial move he enjoyed a degree of reciprocity from Musharraf. Musharraf, it appears had learned that no quick resolution to long-standing and deep-seated differences was possible through political grand-standing. He had resorted to this tactic in the summer of 2001 during the Agra summit only to have hawkish leaders within the Indian cabinet quickly undermine his efforts. Indeed, without Musharraf's willingness to move forward with a substantive and meaningful dialogue Vajpayee's call for a dialogue would have been in vain. In this case, the talks, which appear to have made much progress, unraveled thanks to a domestic/endogenous shock, namely the political agitation within Pakistan that forced Musharraf from office.

In the aftermath of this crisis, after considerable prodding by the United States, some fitful efforts at rapprochement have taken place. For example, in April 2005 the two sides initiated a bus service linking their respective portions of the disputed state of Kashmir (Johnson 2005). The prospects of a breakthrough, however, remain unlikely. Far too many obstacles still lie in the path of any possible rapprochement. The Pakistani military, which received an important reprieve after September 11, 2001, and is now firmly ensconced in ruling the country, does not stand to benefit from a rapprochement with India. Such a settlement would call into question the very substantial resources at its command. At the same time, no regime in India will make significant territorial concessions for reasons that have been already discussed. More to the point, given that India is now on a sustained path of economic growth it can afford to devote substantial resources to its military and simply contain the Kashmir problem. The material and psychic costs of holding on to Kashmir are well within the purview of the Indian state. Accordingly, while it may be willing to pursue a range of confidence-building measures with Pakistan there is no willingness to pursue territorial compromise.[22]

A Dialogue of the Deaf: 2010

In April 2010 on the sidelines of the annual meeting of the South Asian Association for Regional Cooperation, India and Pakistan chose to resume the dialogue that had been suspended in the aftermath of the Bombay terrorist attacks (Sharma 2010). It is widely believed that India chose to resume these discussions. Accordingly, the two foreign secretaries, Nirupama Rao of India and Salman Bashir of Pakistan, met in New Delhi in February 2010. These talks quickly deadlocked as the two sides remained wide apart on critical issues dealing with Kashmir and terrorism. However, they agreed to continue the dialogue (Husain 2010; see also Magnier 2010).

Accordingly, in July 2010, the Indian minister for external affairs, S. M. Krishna, traveled to Islamabad to resume discussions. This meeting, however, proved to be an utter contretemps. Shortly before Krishna's departure for Islamabad, in response to a direct question from a reporter, the Indian home secretary, G. K. Pillai, publicly stated that during his interrogation in Chicago, David Coleman Headley, a Pakistani-American charged in connection with the Bombay (Mumbai) terrorist attacks, had revealed that Pakistan's Inter-Services Intelligence Directorate, had helped coordinate the attack.[23]

Not surprisingly, the talks ended in a complete deadlock with mutual recriminations (Ganguly 2010). The popular explanation for the breakdown of the talks suggests that Pillai's untimely revelations about Headley undermined the talks (*Hindustan Times* 2010). There may be some truth to this allegation. However, it is also possible to argue that Shah Mehmood Qureshi, Pakistan's foreign minister, did not command the confidence of the all-powerful Pakistani military establishment. At a time when the security establishment under General Kayani found that it could influence Afghanistan's internal politics in light of the prospective U.S. drawdown of forces starting in July 2011, it saw little reason to improve relations with India (Tisdell 2010).

A change-seeking entrepreneur, Prime Minister Manmohan Singh had initiated this dialogue. Unfortunately, his interlocutor, President Asif Ali Zardari of Pakistan, was not an independent actor. The peculiar structure of civil-military relations within Pakistan had hobbled his policy choices. Consequently, his room for maneuver and that of his foreign minister, Shah Mehmood Qureshi, was severely constrained. The entrenched hostility of the Pakistani military establishment toward India makes reciprocity exceedingly difficult.

Conflict Escalation and De-Escalation in the Indo-Pakistani Rivalry

Why have multilateral, third party, and bilateral initiatives failed to resolve this conflict? Why have both endogenous and exogenous shocks failed to dramatically alter the course of this rivalry in a positive fashion? The answer to these questions will necessarily be partial. One possible way to address them may be to tease out some general propositions about this rivalry from the complex historical record.

First, the rivalry involves two states with markedly different regional agendas. India, for the most part, has been and remains a status quo power. Even in the aftermath of the war in 1971, India did not engage in any attempt at territorial aggrandizement. It did contribute to the breakup of Pakistan and the creation of Bangladesh. However, it did not in any way seek to augment its territorial status. Pakistan, in contrast, has consistently been a revisionist power. Despite grave asymmetries of power, it has repeatedly sought to change the territorial status quo in South Asia since 1947. Even after the overt nuclearization of the region in 1998, it attempted to alter the status quo in Kashmir in 1999. Amid the ongoing discussions of 2005, Pakistani interlocutors continue to keep Kashmir at the forefront in all of their discussions with India. Pakistan's obsession about Kashmir is certainly one of the factors that has helped keep the rivalry alive. As Steve Cohen, a noted U.S. observer of Pakistan, has written (Cohen 2004): "Over the years, Kashmir has become part of the Pakistani identity—at least, of those Pakistanis who focus on strategic and security issues, notably the army—and it raises deep passions and emotions, especially among the large Kashmiri population in important Pakistani cities."

If the defense of secularism is no longer as critical for India, then what explains India's unwillingness to accept a territorial compromise? The answer is complex. In considerable part it is because Indian elites fear an internal domino effect. If Kashmir or a part thereof is allowed to secede from the Indian union, then a powerful demonstration effect could be set into motion. Other disaffected groups in key states may see the Kashmiri secession as a precedent for their own.[24]

A second explanation for this enduring strategic rivalry lies in state structure. The domestic features of the Pakistani state, where the military establishment is *primus inter pares*, makes the resolution of this conflict exceedingly difficult. Embellishing the putative threat from India justifies substantial military expenditures in Pakistan and enables the military to retain its corporate privileges. To this end, it routinely fans the flames of hypernationalist propaganda and seeks to maintain an image of India's seemingly implacable hostility toward the very existence of the Pakistani state. Civilian regimes have not performed markedly better on this score.

Since they have existed at the sufferance of the military their ability to shift the terms of popular and even elite discourse within Pakistan has been limited (Jones 2002).

More to the point, not a single civilian regime in Pakistan has been allowed to sustain a full term in office. Consequently, democratic consolidation has failed to take place within Pakistan. Fledgling democratic regimes, often seeking quick populist pathways to bolstering their legitimacy, have simply sought to outbid the military in their dealings with India. As a consequence, neither weak civilian regimes nor well-entrenched military regimes have evinced much interest in seeking rapprochement with India.

The depth and intensity of this rivalry precludes most possibilities of its swift resolution. In all likelihood the rivalry will continue for the foreseeable future. In the longer run, two key factors will determine the dimensions of the rivalry. First, the rivalry will lose its significance as India continues to grow at much higher rates, becomes more closely integrated into the global economy, and wields a much more powerful and sophisticated military. Second, it will also abate as India manages to limit the scope of internal conflicts that are susceptible to external manipulation. Both factors should work in tandem. The growing military and economic gap between the two states should limit Pakistan's capacity and propensity to provoke and needle India. Simultaneously, India's ability to forestall, manage, and contain internal conflicts will hobble the opportunities for Pakistani involvement.

Of course, the gap between the two states is not exactly a new factor. Capability asymmetry has hardly constrained Pakistan from provoking India to date. Nor is Indian success in managing its internal conflicts something to be taken for granted. With the passage of time, while the two states may not forge cordial relations, the rivalry could cease to matter to the extent that India's substantially greater military prowess, its economic stature, and its political stability prevails. Then again, there is no guarantee how long this passage of time will take.

But where do external shocks, third-party pressures, and change-seeking entrepreneurs fit into this picture? Despite the presence of these factors, there has been only one important time span (1972 to 1979) when the rivalry saw significant de-escalation. However, even during this period the underlying sources of discord remained unabated. Intransigence and a lack of reinforcement from the adversary as well as other exogenous shocks brought this period of de-escalation to a close.

Table 6.1. Shocks in the Indo-Pakistani Rivalry, 1947–2010

Shift in external threat	*Shift in competitive ability*	*Shift in regime orientations/ strategies*	*Economic resource Crisis*
	War defeats, but the implications have rarely been acknowledged (Pakistan)		Persistent but not sufficiently acute (Pakistan)

What then does the Indo-Pakistani rivalry tell us about our expectancy theory? Table 6.1 suggests, not unlike the Israeli-Syrian case, that shocks—which have certainly been present in the relationship—do not appear to be critical to bringing about rivalry de-escalation/termination. Or, perhaps one should infer that this rivalry has not experienced the appropriate type of shock so far. Yet the Indo-Pakistani case is much like the Israeli-Syrian case in the sense that the weaker member of the pair has been defeated substantially and repeatedly. One might think defeat in war would be as great a shock as is imaginable, subject, that is, to one important caveat. Defeat in war once meant (as recently as 1945) that the losing side lost much if not all of its territory and independence. Both Syria and Pakistan have lost territory, but they have not lost their independence. War defeats can be less shocking than they might otherwise be if the full impact of losing a war is constrained by international norms and third parties.

Just how important this might be in the Indo-Pakistani case is difficult to assess. Equally important, though, we find no links between shocks and expectancy revision. The shocks are there, but their impact is usually to make the rivalry relationship more intense. No expectancy revision has been demonstrated to date. Both sides continue to regard the other as threatening. The most appropriate sort of shock is one that leads directly to expectancy revision and, for various reasons as recounted above (for instance, misperceptions and the nature of Pakistan's political elite), that has yet to come about.

Some of the other factors are not entirely absent in Table 6.2. Third-party pressures have been evident from time to time. Some acts of reciprocity, such as reducing barriers to travel or exchanging prisoners, have been seen as well. But these factors do not carry sufficient weight, in and of themselves, to bring about a substantial de-escalation. The relative absence of

Table 6.2. Modeling Outcomes for the Indo-Pakistani Rivalry

Variable	*1962–1963*	*1966*	*1972*	*1980*	*1989*	*1999*	*2003–2007*	*2010*
Shocks	Present	Present	Present	Present	Present	Present	Absent	Absent
Expectancy revision	Absent	Absent	Some	Absent	Absent	Absent	Absent	Absent
Consolidated and new policy entrepreneur	Absent	Absent	Absent	Absent	Absent	Absent	Present	Present
Third-party Pressure	Present	Present	Absent	Absent	Absent	Absent	Absent	Absent
Reciprocity	Absent	Some	Some	Absent	Absent	Absent	Absent	Absent
Reinforcement	Absent	Absent	Some	Absent	Absent	Absent	Absent	Absent
Outcome	No change	No change	Some change	No change	No change	No change	No change	No change

new change entrepreneurs in this rivalry makes it rather difficult to assess its importance. That evaluation will require more cases.

Significant de-escalation of the Indo-Pakistani rivalry has only taken place in the face of shocks that have left the Pakistani military establishment weakened. Consequently, the rivalry lessened significantly in the aftermath of the war in 1971, which had left the military in morally discredited and materially crippled. The military's intransigence toward India, which helps sustain its extraordinary privileges, cannot be easily overcome. One of two possibilities will end this rivalry: 1) the Pakistani state and especially the military apparatus will inexorably come to the realization that they cannot compete with India and that it thereby makes sense to abandon a policy of unremitting hostility (Ganguly 2006); or 2) a powerful third party, which has the incentive and the capabilities to alter the internal structure of the Pakistani state, will diminish the overweening role of the security establishment. In the absence of these two conditions it is unlikely that there will be much abatement of the Indo-Pakistani rivalry in the foreseeable future.

CHAPTER 7

Other Eurasian Rivalries and Their Interdependence

What we have been describing in earlier chapters resembles the following account. A dyad or pair of states at time t is characterized by a relationship described as rivalry that, in turn, is predicated on a mutual set of expectations about each side's adversary and the ability to deal with the adversary. The consequent rivalry behavior reflects a routine of strategies developed by both sides to deal with the problems posed by their opponent. Given the right sort of conditions, such as the confluence of shocks, consolidated policy entrepreneurs, reciprocity, and reinforcement, expectational revision may occur. Some change in strategies should follow, although not necessarily immediately. If enough expectational revision and strategic change occurs, then the dyad at time $t+1$ may either no longer be in a state of rivalry or, at the least, the rivalry might have de-escalated significantly.

The real world, of course, is more complicated than a single dyad.[1] In a world of literally thousands of dyads, a change in one dyad might not make much difference to the overall population of strategies. Then again, it depends on which dyads change (evolve). The Sino-Soviet case is a good example. All dyads are not of equal weight in shaping the landscape of international politics. Imagine a landscape characterized by a collection of mountains, hills, and bumps. Most interstate dyads can be compared to the bumps. A few are hills. A very select few constitute the mountains of international politics. By global standards, there may be some reason to question whether the Sino-Soviet case was a mountain. It certainly was not the tallest mountain in the post–World War II landscape. But the Sino-Soviet dyad was certainly more than a bump and probably more than a hill. By Asian standards (as opposed to global standards), it was clearly a mountain—just as the Russo-Chinese relationship remains one of several mountains in the Asian international political landscape.[2]

The Sino-Soviet rivalry had importance because of the large populations and military capability of the two states involved. But the rivalry also had importance because it was interdependent with (and within) a wider set of rivalries and relationships. How the Soviet Union related and Russia relates to China and vice versa has had and continues to have implications for a number of other dyads. It certainly mattered to the most important major power triangle (the United States, the Soviet Union, and China). Just how it mattered, and whether it still matters, is the subject of debate. Yet Soviet-U.S. interactions first encouraged Chinese conflict with the Soviet Union and then encouraged Chinese rapprochement with the United States. Ironically, it may later have also encouraged the de-escalation of the Sino-Soviet rivalry.

However the major power triangle works (or has worked), it is clear that Sino-Soviet conflict has influenced the way in which Chinese-Vietnamese and Sino-Indian rivalries operate. Without Soviet support, Vietnam and India have demonstrated less overt hostility toward China. The collapse of the Soviet Union had to have been a shock for Vietnamese and Indian decision makers. These altered relationships could also be expected to have consequences for rivalries "once removed," as in the cases of Vietnam-Thailand and India-Pakistan. Other East Asian rivalries, China-Taiwan and the two Koreas, and even nonstrategic rivalries, China-Japan and Russia-Japan, should also be expected to feel the reverberations of the change in the landscape (the reduction of the Sino-Soviet, or Sino-Russian, mountain). And since the collapse of the Soviet Union was closely related to Afghani turmoil, it is not too far-fetched to relate the de-escalation of Sino-Soviet rivalry, albeit indirectly, to the former rivalry between Pakistan and Afghanistan and perhaps even the short-lived one between Pakistan and Iran.

The Sino-Soviet rivalry, then, was not your average run-of-the-mill protracted conflict. It had implications for relationships in the Asian major power triangle, as well as for relationships in Northeast and Southeast Asia (both the continental and the maritime subregions), and South Asia. No doubt, one could make a case for extending the reach of the rivalry to the African continent as well. The de-escalation of the Sino-Soviet rivalry had significant reverberations throughout the system not unlike if not quite on the same order as the de-escalation of the Soviet-U.S. rivalry.

In this chapter, we wish to highlight the notion that rivalry termination—or, at least, some rivalry terminations—can contribute to the rivalry de-escalation process. We are not abandoning our focus on shocks, expectations, reciprocity, and reinforcement. But we do feel the need to deviate from the standard focus on each rivalry in its own terms. That rivalries are not islands unto themselves, as John Dunne might have said, is one of the biggest risks of engaging in rivalry analyses. Rivalries tend to be embedded in larger networks of rivalries.[3] Changes wrought in rivalries that are central to a specific field of adversarial relationships can constitute shocks that reverberate throughout the network. In other words, rivalry de-escalation can contribute to further rivalry de-escalation. We need to be careful, therefore, not to ignore this contextual dimension of the termination process.

The Sino-U.S. and Sino-Soviet Rivalries

It is easily possible to argue that three of the most significant rivalries of the past half-century or so have been the three involving the United States, the Soviet Union, and China. Their significance is guaranteed by the major power status of the rivals and the implications of their rivalry relationships for other states in the system.

In this chapter, we look at two of the three rivalries—the ones involving Chinese confrontations with the United States and the Soviet Union. We do not ignore the rivalry between the United States and the Soviet Union because it was somehow less important than the two that we are focusing upon. Far from it, but the U.S.-Soviet rivalry is more complex. Its implications were also more global in scope than the two rivalries in which China was a participant. Thus, we think we can "handle" the de-escalation of the Sino-U.S. and Sino-Soviet rivalries fairly quickly.[4] They certainly overlap one another and each one had implications for the other one. Equally important for our purposes, they also had a number of implications for other rivalries involving weaker powers in Asia. Examining the Sino-U.S. and

Sino-Soviet rivalries therefore sets us up nicely for a subsequent, second order analysis of rivalries between the two Koreas, China and Taiwan, China and Vietnam, and Thailand and Vietnam.

The Sino-U.S. Rivalry

A postwar strategic rivalry between the United States and China, of course, was never inevitable.[5] It does seem, though, that it was certainly highly probable once the U.S.-supported Nationalists (KMT) lost the civil war in 1949 and withdrew to Taiwan.[6] When that occurred, regaining control of Taiwan, whether one regarded it as a logical continuation of the mainland civil war or an exercise in irredentism, became a prime goal of Chinese foreign policy. Decision makers in the United States might have remained aloof while the mainland ultimately established control on both sides of the Taiwan Straits, but the shock of the Korean War sealed the onset of a rivalry relationship between China and the United States. One of the responses of the United States to the North Korean invasion was to move Taiwan into the U.S. Pacific defense perimeter, along with South Korea.[7] The U.S. Seventh Fleet soon began patrolling the Taiwan Strait, initiating an increasing commitment to the defense of the KMT regime in Taiwan. Between the twin and very proximate threats of an overthrow of the North Korean regime and the maintenance of the KMT on Taiwan, the United States assumed the role of China's primary source of military threat in 1950. China, for its part, came to be seen as the leading source of major power threat to the U.S. position in eastern Eurasia.

These reciprocal circumstances changed little through the 1950s, with Chinese-U.S. crises in 1954–1955 and 1958 over possible Chinese moves on Taiwan, or at least the islands of Quemoy and Matsu between Taiwan and the mainland, and continued into the 1960s. By the late 1950s, however, China's national security problems had become more complex. The Sino-Soviet split had expanded the sources of proximate military threats along the extremely long Chinese-Soviet border. The Chinese chose to adopt a dual adversary stance designating both the United States and the Soviet Union as its primary enemies. Several possibilities of superpower preemptive strikes on the growing Chinese nuclear arsenal occurred in the 1960s. So, too, did the escalation of the U.S. involvement in South Vietnam and the increased probability of renewed combat between the United States and

China. If a drive into North Vietnam by the United States had threatened the existence of the Communist regime there, then a replay of the intervention of China in 1950 was a conceivable development—one that both sides seemed eager to avoid but could not completely rule out.

Several new shocks toward the end of the 1960s (see Table 7.1) altered the political-military landscape in eastern Eurasia.[8] Ross (2001:15) argues that a principal reason for the prolonged U.S.-China rivalry was "each side's exaggerated appraisal of its counterpart's threat and its own ability [to countervail]." The multiple shocks served to diminish some threat perceptions and the confidence that the antagonists had the capability to cope with their policy problems. The first shock was the Tet-driven realization by the United States that it was losing its war in Southeast Asia. By 1968, President Johnson had decided not to run for reelection, domestic support for the war was declining, and a U.S. de-escalation of the war seemed increasingly likely. In 1969, the newly elected president, Richard Nixon, began to speak of U.S. allies in Asia doing more for their own defense while the United States would do less. Some reduction of the U.S. presence in Asia seemed likely.

From a Chinese perspective, the diminishing military threat in the south was paralleled by increasing threat in the north. In late 1968, the Brezhnev Doctrine was announced to justify intervention in Czechoslovakia. In March 1969, the border clashes at Zhenbao/Damansky Island escalated into a potential Sino-Soviet war. Since the Chinese are usually given credit for initiating this crisis—presumably to deter a possible Czech-like Soviet attack in the future—the specific crisis was probably a greater shock for the Soviets than the Chinese.[9] Yet it was also clear to the Chinese who their primary military threat now was. While the immediate threat of war receded, due in part to increased Chinese cooperation over border negotiations, Soviet preparations for a possible invasion somewhere along the border expanded.[10]

A third shock had already come in the form of turmoil in domestic Chinese politics. The Cultural Revolution in 1967 and 1968 left the impression of an uncontrolled political system unable to mount much of a threat outside its own borders but quite capable of weakening Chinese capabilities and further isolating the Chinese from the outside world.[11] By 1969, the Chinese were facing differential external threats—diminishing in the south but increasing in the north—and few sources of assistance in sight. As Mao Zedong reconsolidated central control within China, the wisdom of maintaining the

Table 7.1. A Selected Chronology of Sino-U.S. Interactions, 1944–1979

Date	*Event*
October 1944	The United States reverts to exclusive patronage to Nationalists (KMT) after a brief period of supporting both Communists (CCP) and the KMT in resistance to Japanese occupation of China.
November 1944	The U.S. ambassador attempts to broker a CCP-KMT coalition that falls apart when Communists are told that they must disarm.
January 1945	Mao attempts unsuccessfully to arrange a meeting with Roosevelt to explain the Communist position and the U.S. ambassador's bias in favor of the KMT.
February 1945	At Yalta, the Soviets agree to support the KMT exclusively in return for assets in Manchuria and control of Outer Mongolia.
April 1945	The U.S. ambassador to China denounces the CCP as the main source of disorder in China; Mao warns the CCP Congress that the United States and the KMT would attack the CCP as soon as Japan surrenders.
May 1945	Truman's foreign policy advisers view Chinese Communism as a prelude to Soviet domination in Asia.
August 1945 onward	The CCP and the KMT compete to control territory surrendered by the Japanese; the United States insists that the Japanese must surrender only to KMT forces and deploys 60,000 troops to assist the process; Soviet troops in Manchuria begin cooperating with the CCP.
January 1947	The Marshall Mission to remove Japanese, U.S., and Soviet forces from China and to arrange a compromise between the CCP and the KMT succeeds on the first goal and fails on the second.
1949	The CCP is in control of most of mainland China with the KMT leadership having fled to Taiwan; the United States rules out intervening in the Chinese civil war or defending Taiwan; the People's Republic of China (PRC) is proclaimed in October in Beijing.
December 1949–February 1950	Mao is in Moscow negotiating Soviet assistance and alliance.
January 1950	The PRC closes diplomatic offices of states that have not yet recognized the PRC, thereby ending official contacts between the United States and China; the United States and China are already supporting opposing sides in the fighting in Vietnam; the United States announces an end to military aid to the KMT on Taiwan, preferring to contain Chinese expansion with a defensive arc from Japan to the Philippines to Southeast Asia.

Date	Event
June 1950	North Korea attacks South Korea; the United States commits to defending South Korea and Taiwan, and providing increased assistance to the French in Indochina; in July, MacArthur urges U.S. assistance for KMT attacks on southern China.
October 1950	Chinese troops intervene as it appears that the U.S.-led coalition will destroy North Korean forces and expand southern control throughout the Korean peninsula.
July 1953	The armistice in the stalemated Korean War is signed; the Eisenhower administration continues hard-line containment of China with trade controls and security pacts with Japan, South Korea, Thailand, the Philippines, South Vietnam, and Taiwan.
September 1954	The PRC begins shelling the KMT-held islands of Quemoy and Matsu.
December 1954	The United States offers Taiwan a mutual defense pact that limits the protection of off-shore islands and requires the KMT to consult with the United States before attacking the PRC.
April 1955	The PRC ceases shelling the off-shore islands, leading to intermittent ambassadorial talks between the United States and China in Geneva and Warsaw.
August 1958	The shelling of Quemoy and Matsu resumes; the United States responds by increasing its military presence and veiled threat of nuclear weapons.
Early 1960s	The Kennedy administration views the Chinese threat as manifested primarily in Chinese support for guerrilla warfare in Asia and Africa; U.S. support for Vietnam's defense is expanded and support for Taiwan is also increased.
Mid- to late 1960s	The fighting in Vietnam is seen by both U.S. and Chinese sides as a proxy contest over containment of China, with as many as 328,000 Chinese troops in noncombat roles in North Vietnam between 1965 and 1968.
Late 1968	The Tet Offensive suggests that U.S. efforts in Vietnam are failing; Johnson decides not to run for reelection; his successor, Nixon, hints at a new approach to U.S.-Chinese relations.
March 1969	Sino-Soviet fighting escalates over control of Zhenbao/ Damansky Island.
1969	The United States seeks Soviet and then Chinese assistance in winding down the Vietnam War.

(continued)

Table 7.1 (Continued)

Date	*Event*
Late 1969	The United States offers to end U.S. naval patrols of the Taiwan Straits to facilitate talks with China.
April 1970	U.S. troops invade Cambodia, setting back U.S.-Chinese negotiations.
Summer 1970–September 1971	The internal power struggle between Mao and Lin Biao in part over policy toward the United States culminates in the death of Lin Biao.
April 1971	U.S. ping pong team visits China; later in the month, the United States is invited to have Nixon or Kissinger visit Beijing without requiring that U.S. forces be withdrawn from Taiwan.
July 1971	Kissinger and Zhou hold secret talks in Beijing.
October 1971	Kissinger visits China for a second time.
February 1972	Nixon visits China with the United States acknowledging the concept of one China.
March 1973	Kissinger informs the Chinese that the United States would cut ties to Taiwan and normalize relations with China, but the Watergate scandal soon derails the Nixon administration; the successor Ford administration postpones normalization to better deal with conservative criticism—paralleled in part by political succession problems in China (Zhou and Mao die in 1976).
January 1979	After extended negotiations, the United States cuts ties to Taiwan and recognizes the PRC in part to put more pressure on the Soviet Union; China wants recognition to facilitate the inflow of western capital and U.S. acquiescence in attack on Vietnam.

Source: Based primarily and selectively on the discussion of events in Schaller 2002.

dual adversary strategy seemed ripe for reassessment, even if it was not obvious in which direction a change in policy should be oriented. One faction, led by Lin Biao, favored improving relations with the Soviet Union; another, with Zhou Enlai as its lead spokesman, favored exploring possibilities with the United States. The Lin Biao faction ultimately lost the domestic struggle for policy control and its leader died while attempting to flee to Moscow (Garver 1982: 124–140).

It helped that U.S. decision makers were open or opening to the idea of rapprochement with China. There had been little prospect of that happening during the Truman, Eisenhower, or Kennedy administrations (Foot 1995: 114–142). Some receptiveness emerged too late in the Johnson administration to be useful (Ross 2001: 13; Schulzinger 2001). The irony is that it was Richard Nixon, the old anti-Communist warrior, who had begun to contemplate a different approach to China by the mid-1960s. Initially, his national security advisor, Henry Kissinger, was aghast at the idea of improving relations with China but eventually came around (Mann 1999: 17–19). But the combat between the Soviet Union and China in March 1969 appears to have had the effect of convincing U.S. decision makers that the split between the two Communist states could be utilized to U.S. advantage (Schaller 2002: 168).

Even though the incumbent U.S. president was unusually open to the idea of rapprochement with China, U.S.-Chinese relations were relegated to a secondary priority.[12] First priority was given to U.S.-Soviet relations.[13] By the late 1960s, U.S.-Soviet relations had improved over what they had been much of the time prior to 1962. Yet Soviet nuclear weapons were in the process of developing parity with those of the United States. A Soviet quantitative lead in warheads was thought to be conceivable. Decision makers in the United States sought ways to encourage further gains in U.S.-Soviet détente and arms control negotiations. Another way in which the Soviets might be helpful was in encouraging the North Vietnamese to negotiate a winding down of the Southeast Asian conflict that would facilitate a U.S. withdrawal. Better relations with the Chinese, it was thought, would encourage, perhaps even ensure, greater cooperation from the Soviets.[14]

Signaling began between the United States and China in 1968 and continued up to and including the visit of Nixon to China in early 1972.[15] The main negotiating problem was a pairing of the Vietnamese-Taiwanese issues. The Chinese wanted the United States to abandon its protection of Taiwan. The United States wanted the Chinese to encourage the North Vietnamese to negotiate with the United States. It is not fully clear how far U.S. spokesmen went verbally on the issue of Taiwan, but the U.S. willingness to compromise stopped short of fully abandoning Taiwan immediately.[16] Chinese pressure on the North Vietnamese proved unsuccessful. The signaling was also slowed and made more difficult by the movement of U.S. forces into Cambodia, thus further escalating the scope of the Indochinese fighting. From the Chinese perspective, it revived the possibility of a southern threat and muddied the implications for a U.S. withdrawal from Southeast

Asia. The Chinese also had an internal distraction in the form of factional infighting that contributed to slowing the rapprochement process (Schaller 2002: 170).

The Sino-Soviet Rivalry

The immediate question is whether this perspective helps to account for the ups and downs of Sino-Soviet relations since 1969 when the rift between these two countries peaked with relatively large-scale border clashes.[17] Since that time, China first moved closer to the United States in the 1970s in search of some external protection and then attempted to distance itself from both the Soviet Union and the United States in the 1980s. In the 1990s and through the first decade of the 2000s, China and Russia have given the appearance of moving back toward a more cooperative relationship that may or may not stop short of recreating the alliance of the early 1950s. During this same period, the Chinese initially upgraded the Soviet Union to the role of its principal rival (supplanting the United States), de-escalated its rivalry with the United States, and then proceeded slowly to de-escalate its rivalry with the Soviet Union. The Soviet Union is no more, but new versions of some of the rivalries among the United States, Russia, and China have reemerged. While it might be most appropriate to tackle the simultaneous de-escalation of the rivalries between the United States and the Soviet Union, the United States and China, and China and the Soviet Union, the scope of such a task is too great for our immediate purposes. Focusing on two of the three, the Sino-U.S. and Sino-Soviet rivalry de-escalations, will need to suffice.

The Sino-Soviet dispute had a reasonably long and complicated history. One might easily start with the initial Sino-Russian clashes that date back to Russian expansion into Siberia in the seventeenth century. The Russian encroachments on Chinese territory in the nineteenth century were part of the same historical process. The most recent phase of Sino-Russian conflict dates back to the brief interlude between the late 1940s and the 1950s. The alliance of the Soviet Union and China appeared to represent a Eurasian monolith threatening to expand to the west, the east, and the south. In fact, Sino-Soviet cooperation was very short-lived. After the death of Stalin and the emergence of a new Soviet approach to dealing with the West, Mao and China began to distance themselves from the Khruschev-led, Soviet preference for coexistence. On the one hand, the dispute was about Marxist ideology and policy in the East-West struggle, with the Chinese holding out for a harder line. This dimension turned into a competition for leadership within the Communist international system. On the other hand, the dispute was also about dyadic subordination. The Chinese resented playing the role of weaker partner. China needed technology, financial assistance, arms, and protection. The Soviet Union was prepared to offer these things but at a price that included acknowledging Soviet superiority in the relationship.

The Chinese began to move away from accepting the Soviet lead within the Communist subsystem after 1956. By the early 1960s, the rift was out in the open. By the end of the 1960s, the probability of war between the two states seemed high. In reaction, the Chinese sought a new relationship with the state that had formerly presided as the principal Chinese adversary, the United States. For a China still seeking protection, technology, financial assistance, and arms, a rapprochement with the United States seemed the answer. Such an arrangement served U.S. purposes in Southeast Asia and in terms of its Soviet containment objectives. Yet as the Sino-U.S. courtship seemed to be intensifying, the perceived threat to the Chinese from the Soviet Union was receding. This change facilitated a change in Chinese strategy that involved maintaining some distance from both the United States and the Soviet Union while, at the same time, continuing to negotiate/cooperate with both superpowers on a selective basis.

The Soviet Union appeared to be ready to negotiate at least some of its differences with China, albeit on its own terms, as early as 1969/1970. Several Soviet initiatives were advanced in the form of proposals for a nonaggression pact, presumably with the intention of lessening the high level of threat perceived by Chinese decision makers. The Chinese remained unresponsive throughout most of the 1970s. China's probes toward patching up its differences with the Soviet Union began in 1979 (Chi Su 1993: 49). The initial Soviet response was cautious, preferring to focus first on nonpolitical exchanges of students and delegations. In particular, Soviet decision makers were reluctant to discuss what they regarded as third-party situations in Cambodia and Afghanistan, while Chinese decision makers posed these issues as major stumbling blocks (the "Three Obstacles," which included troop levels and boundary concerns along their mutual border) to further progress.

Soviet receptivity began to be manifested more clearly in the early 1980s even though it may have reflected a compromise between hard-liners and moderates. After the ascension of Gorbachev in 1985, Soviet receptivity became even more clear. By 1985, the Chinese were also wavering a bit on the extent to which the Three Obstacles were genuine barriers. Alternatively put, the Chinese had signaled that compromise on one of the three, the Cambodian issue, might suffice. Yet the two states seemed unable to come to terms until 1989—a full decade after the initiation of overtures toward de-escalation from both sides. Why did the process take so long to reach fruition when both sides appeared interested in reaching some form of accommodation?

Table 7.2 provides some sense of the flow of interaction between the Chinese and the Soviets across three decades. The reported interactions do not represent a systematic inventory of interactions. Instead, they constitute events reported in the literature on the Sino-Soviet rapprochement and thus should be viewed as a form of purposive sampling undertaken by informed specialists on the subject.

The main question to address is the relation between shocks and changes in expectations. The Chinese downgraded their perceptions of the likelihood of a Soviet military attack in 1979 prior to the Chinese overtures for reduced tensions (Shambaugh 1994: 204; Robinson 1994: 572).[18] Apparently, the argument was that if the Soviets were going to attack, they would have done so already. Since they had not done so, the probability of an attack was increasingly less likely. It is not clear that Soviet decision makers ever anticipated a Chinese attack per se. Their security concerns were always more long term in anticipation of increased conflict with a large Chinese population needing more space in Northeast Asia and/or attempting to regain northern territory once part of the Chinese Empire. These concerns do not appear to have completely vanished from the Russian policy scene.

Both sides moved toward a posture that emphasized the need for a peaceful external environment while maximal energies could be focused on domestic reforms and rebuilding. China moved first, in 1979, as articulated in Deng's Four Modernizations. The Soviet Union followed after 1985 with Gorbachev's perestroika movement. This convergence (Legvold 1993: 81) comes under the label of "domestic preferences emphasizing domestic concerns".

While this convergence was crucial, it also had two implications. Soviet perestroika stemmed from an interpretation that Soviet capabilities had fallen behind those of the West and that radical reform was necessary to stay in the competition. The Chinese agreed with this assessment. At one point, when their own sense of Soviet threat was acute, Chinese decision makers portrayed the Soviet Union as surpassing U.S. capabilities. In the

Table 7.2. A Chronology of Selected Sino-Soviet Interactions, 1960s–1990s

Date	*Event*	*Outcome*
February–August 1964	Border talks in Beijing	No agreement
June 1969		Soviet Union proposes "no first attack" policy
June–August 1969	River navigation commission	Some agreement
October–December 1969	Border talks in Beijing	No agreement
January–April 1970	Border talks in Beijing	Soviet Union conciliatory; no agreement
February 1970		Soviet Union proposes both sides make no territorial claims and accept status quo
May 1970	River navigation commission	Some agreement
January–summer 1971	Border talks in Beijing	Soviet Union offers concessions, including nonaggression pact; no agreement
December 1971–March 1972	River navigation commission	No agreement
March–July 1972	Border talks in Beijing	No agreement
January–March 1973	River navigation commission	No agreement
March–June 1973	Border talks in Beijing	Soviet Union offers nonaggression pact again; no agreement
February–March 1974	River navigation commission	No agreement
June–August 1974	Border talks in Beijing	China conciliatory if Soviet Union will agree to mutual withdrawal from disputed areas; no agreement
October 1974		Soviet Union offers nonaggression pact again
November 1974		China offers to accept nonaggression pact if coupled to mutual withdrawal from disputed territories
February–May 1976	Border talks in Beijing	No agreement
January 1976		Death of Zhou Enlai
April 1976		Ouster of Deng Xiaoping
September 1976		Death of Mao Zedung
October 1976		Ouster of Gang of Four
November 1976–February 1977	Border talks in Beijing	Soviet Union conciliatory; no agreement

(*continued*)

Table 7.2 (Continued)

Date	*Event*	*Outcome*
July 1977		Deng Xiaoping assumes power; talks on river navigation resume
February 1978		Soviet Union proposes joint statement of principles
May–June 1978	Border talks in Beijing	Agreement to suspend talks
February–March 1979	River navigation commission	Some agreement
April 1979		China proposes general negotiations after declining to renew alliance of 1950
September–November 1979	Normalization talks in Moscow	China establishes the Three Obstacles to normalization (troop reductions along border, withdrawal from Outer Mongolia, and an end of support for Vietnam in Cambodia); Soviet Union proposes expansion of nonpolitical interactions
March 1980	River navigation commission	No outcome
February–March 1981	River navigation commission	Some agreement
August 1981		Soviet Union proposes confidence-building measures in Far East to China
September 1981		Soviet Union proposes resumption of border talks
December 1981		Soviet Union proposes nonpolitical exchanges
February–March 1982	River navigation commission	Some agreement
March 1982		Tashkent speech signaling Soviet willingness to improve relations
April 1982		Agreement to resume border trade
June 1982		Soviet Union pledges no first use of nuclear weapons
July–August 1982		Soviet Union proposes resumption of Soviet-Chinese negotiations without preconditions; China relaxes Soviet troop withdrawal precondition
October 1982	Normalization talks in Beijing	Some agreement on border issues; China expresses the Three Obstacles
November 1982		Death of Brezhnev
March 1983	Normalization talks in Moscow	Soviet Union offers nonaggression pact; China invokes the Three Obstacles
October 1983	Normalization talks in Moscow	Agreement to increase nonpolitical interactions

Date	*Event*	*Outcome*
February 1984		Death of Andropov
March 1984	Normalization talks in Moscow	Disagreement focuses on Sino-Vietnamese conflict
October 1984	Normalization talks in Moscow	Agreement on nonpolitical interactions
March 1985		Death of Chernenko; Gorbachev assumes power
April 1985		China notes that if the Three Obstacles are too difficult to eliminate at once, the Soviet Union could start with one (Vietnamese withdrawal from Cambodia)
October 1985	Normalization talks in Beijing	Little progress
March 1986	River navigation commission	Some agreement
March–April 1986	Normalization talks in Moscow	China reinvokes Three Obstacles but agrees to resume border talks
July 1986		Vladivostok speech addressing Three Obstacles in conciliatory way; China responds by toning down verbal attacks
October 1986	Normalization talks in Beijing	Both sides hint at compromise but deadlock on Cambodian issue
April 1987	Normalization talks in Moscow	Agreement on consular arrangement
October 1987	Normalization talks in Beijing	No agreement on Cambodian issue
January 1988	Normalization talks in Moscow	Begin discussion of troop withdrawals from border
August–September 1988	Normalization in Beijing	Some agreement on Cambodian issue
September 1988		Kranoyarsk speech—Soviet states prepared to accept confidence-building and other arrangements and discussions
February 1989		China compromises on Cambodian issue; cessation of aid to Vietnam no longer demanded
April 1989		Vietnam announces intention to withdraw from Cambodia
May 1989	Summit meeting in Beijing	Formalizing end of Sino-Soviet dispute

(*continued*)

Table 7.2 (Continued)

Date	*Event*	*Outcome*
April 1990	Summit meeting in Moscow	Agreement on border demilitarization
December 1992	Summit meeting in Beijing	Thirteen agreements signed
September 1994	Summit meeting in Moscow	"Constructive Partnership" rhetoric with agreement not to target nuclear weapons on each other
April 1996	Summit meeting in Beijing	"Strategic Partnership" rhetoric
April 1997		Agreement on demilitarization

Sources: Based primarily on information found in Hart 1987 and Dittmer 1992 and supplemented by Nelson 1989; Legvold 1993; Ellison 1993; Yahuda 1993; Chi Su 1993; Tow 1994; Anderson 1997; and Harada 1997.

1980s, that assessment was scaled back to a view that the United States had renewed its own capabilities and had pulled ahead (Dittmer 1992: 231). A weakened Soviet Union not only meant less threat to China, it also meant that Chinese capabilities relative to those of the Soviet Union were no longer so asymmetrical. Chinese relative weakness was reduced by Soviet weakness. To the extent that Chinese decision makers had been reluctant to negotiate with the Soviet Union prior to 1979 from a position of relative dyadic weakness (see Chi Su 1993: 49 and Goldstein 1994: 241), this attribute of the bargaining situation had been altered substantially (that is it represented a change in relative economic capabilities).

Finally, part of the ideological conflict between China and the Soviet Union had been over tactics and strategies concerning East-West relations and domestic economic policies. China under Mao had wanted a harder line to be taken with the West than Soviet ideas about coexistence encouraged. China under Deng sought substantial reform in domestic economic policies. Since the Chinese had worked out a working arrangement with the West on their own and since the Soviets were moving toward domestic reforms that were not unlike those being pursued in China, Sino-Soviet disputes over internal and external doctrine had been ameliorated considerably. Thus, in a very general sense, changes had occurred in adversarial leadership and institutions that reduced goal conflict.

Multiple shocks may seem to be overdetermining the outcome, though they were less than sufficient. Policy entrepreneurs on both sides were needed

(Deng and Gorbachev)—a subject to which we will return shortly. More than just minimal reciprocity was also needed. The Soviet Union had begun to address the Three Obstacles raised by the Chinese as necessary for more cooperative relations. Soviet troops were being withdrawn from the border with China and disagreements about boundaries were being resolved. Nevertheless, some movement on the Cambodian or Kampuchean issue was needed and eventually forthcoming in 1989 more or less coterminously with the admission of Soviet failure in Afghanistan. Shortly after the Chinese learned that the Vietnamese would be withdrawing from Cambodia, the Sino-Soviet rift was declared dead in a Beijing summit meeting between the two former disputants.

One observation that comes across very clearly in scanning the interaction flow is that mere changes in policy entrepreneurs have not been sufficient in altering the Sino-Soviet relationship. Zhou, Mao, Deng, Brezhnev, Andropov, Chernenko, and Gorbachev have come and gone since the mid-1970s. The Soviets thought the death of Mao might facilitate the quick restoration of good relations between the two states (Ellison 1993: 96), but it did not. Andropov and Gorbachev were thought to be especially open to accommodating Chinese demands, but neither was immediately successful. Deng also pursued some form of de-escalation with the Soviets, but success eluded him for a number of years.

Leadership changes may have delayed a Sino-Soviet accommodation almost as much as they contributed to change. The Soviet Union experienced several changes in leadership between 1982 and 1985 that could not have been conducive to negotiating a higher level of cooperation or a change in the relationship. While state funerals offered opportunities for low-key negotiations, they also suggested high levels of uncertainty about the likely direction of future policies. One important reason is traceable to the lag between assuming power and consolidating it. Deng assumed power in 1977 but was not in a position to make overtures to the Soviet Union until 1979. Similarly, Gorbachev came to power in 1985 but did not make his concessionary bid toward China until the following year. Both leaders first had to purge domestic sources of resistance to their proposed changes in policy (on Gorbachev's purges, see Chi Su 1993: 53). Both were ultimately successful in doing so but not to the point that people wedded to older ways of doing things could not return to power in subsequent years. Such consolidation struggles tend to be ongoing processes—as opposed to a one-time suppression of domestic opposition. They can create windows of opportunity for dramatic changes, but not indefinitely. If the de-escalation process stalls or other shocks occur, then the windows can close as positions of influence are resumed by people associated with older ways of doing things.

The role of continued reinforcement is also underscored by the Sino-Soviet case and now the Sino-Russian case. The Soviet Union treated the massacre in Tiananmen Square in 1989 very gingerly. Both the collapse of the Soviet Union and the subsequent failed attempt to restore it might have altered the strategic environment sufficiently to derail the de-escalation of tensions between the two states, but did not. While there appear to be reservations on both sides about how far to push their relationship (Goldstein 1994: 224; Anderson 1997; Scott 2007: 147–150), the rift has clearly not re-emerged, and, verbally at least, the relationship between the two states has become more cooperative in the 1990s and 2000s—rather than less.

There is more than mere reinforcement going on here. The other factors, especially the effects of earlier shocks and policy entrepreneurship, are still influencing the relationship. Neither side has seen any reason to upgrade the perceived threat emanating from its former opponent. Both sides have increased reason to cooperate and even to collude in opposition to U.S. policy in Asia (Nelson 1989: 131; Legvold 1993: 77; Ellison 1993: 101; Scott 2007: 148). Moreover, the former asymmetry between the Soviet Union and China has not returned as might have been the case if Russia had been able to reconstruct a successful political-economic-military foundation for great power in foreign affairs. This has yet to happen. On the contrary, the capability asymmetries have been reversed. China's demographic and economic size is much greater than Russia's. China's future economic trajectory is also much more promising than Russia's. Moreover, both sides generally continue to value a relatively peaceful external environment so that attention can be focused on domestic problems. Both sides also benefit from their increased economic and military interactions—another side of the reinforcement coin.

In accordance with the expectancy theory, then, the confluence of multiple factors (shocks, expectation revision, policy entrepreneurship, reciprocity, and reinforcement) came together to tip the Sino-Soviet relationship from one of intense conflict to one of at least moderate cooperation. It follows that something along the same lines will be required to tip Chinese and Russian expectations, strategies, and behavior back to the types associated with strategic rivalry.

Analyzing the two major power rivalries involving China is characterized by a problem seen in earlier chapters. Both rivalries have been terminated. Consequently, we have no variance in this small sample. Still, it is worth reviewing and comparing the two cases before we proceed to our last cluster of rivalries in the next chapter.

Table 7.3 summarizes the shock information. The Sino-U.S. rivalry termination appears to be a case of classical realpolitiks. One side, China, perceived a greater threat emanating from a third actor while the U.S. threat had been downgraded. The downgrading occurred because it was apparent that U.S. decision makers recognized that the Vietnam War was a losing proposition and were searching for an exit strategy. Moreover, decision makers in the United States no longer viewed the containment of China as part of the U.S. strategy of containment of the Soviet Union. The split between the two Communist powers had reduced the threat of a pan-Eurasian monolith in Western eyes. At the same time, neither side perceived that its own external or internal position was all that solid. Externally, the United States sought retrenchment. Internally, the Vietnam War had led to massive protests and alienation from governmental policies. The Chinese, for their part, were attempting to recover from the domestic turmoil and external isolation wrought by the Cultural Revolution. Both sides, therefore, were literally shocked into being open to doing something different.

In contrast, the nature of the Sino-Soviet shock was less classical and more subtle. One side, the Soviet Union, realized that it was falling behind its U.S. adversary and needed to try different strategies to improve its deteriorating economic position. Reducing tensions with the Soviet Union's main adversaries was part of the strategies of "new thinking." In this case, the

Table 7.3. Shocks in the Sino-U.S. and Sino-Soviet Rivalries

Rivalry	*Shift in external threat*	*Shift in competitive ability*	*Shift in regime orientations/ strategies*	*Economic resource crisis*
Sino-U.S.	Soviet attack threat (China)	Vietnam (United States) Cultural Revolution (China)		
Sino-Soviet		Perestroika (Soviet Union)	Perestroika (Soviet Union)	

shock of falling behind did not have to lead to efforts at tension reduction. One can imagine a scenario in which Soviet decision makers chose to hunker down and try harder—as opposed to conciliation. It did help, though, that there had been intermittently ongoing efforts to at least discuss Sino-Soviet grievances. The mid- to late 1980s, moreover, were at least a decade and half away from the high peak of Sino-Soviet tension in 1969.

With the Soviets willing to address concretely the Chinese-perceived obstacles to meaningful negotiations, especially Afghanistan and Kampuchea, the Chinese perception of Soviet containment and threat had diminished. Add two consolidated policy entrepreneurs on both sides committed to a preference for fewer immediate enemies to facilitate domestic reforms and the opportunity for rivalry de-escalation was seized. Subsequent developments have worked to reinforce the rapprochement by suggesting that the United States remained a greater threat to Russian and Chinese interests than Russia and China did to one another. Thus, despite different circumstances, the profiles summarized in Table 7.4 for the two major power rivalries resemble each other in the abstract. In each case, a combination of most of the six (but not all of the) factors led to rivalry termination.

Nonetheless, the rivalries among China, the Soviet Union/Russia, and the United States have been, at some times more than others, highly interdependent. What shocked inertial tendencies in these rivalries was not always restricted to one rivalry at a time. Other things being equal, when the United States has been seen as a grave threat by both China and the Soviet Union/Russia, the Sino-Soviet/Russian leg of the triangle has tended to be cooperative as long as the two Asian states have not viewed each other as threatening in a significant way. Of course, other things are rarely equal. China and

Table 7.4. Modeling Outcomes for the Sino-U.S. and Sino-Soviet Rivalries

Variable	*Sino-U.S.*	*Sino-Soviet*
Shocks	Present	Present
Expectancy revision	Present	Present
Consolidated and new policy Entrepreneurs	Present	Present
Third-party Pressure	Absent	Absent
Reciprocity	Present	Present
Reinforcement	Present	Present
Outcome	Termination	Termination

the Soviet Union did for a time see each other as significant threats. More importantly, the relative positions of the three states have not remained constant with two of the three experiencing relative decline and the third enjoying an ascending trajectory. Foreign policy successes and failures have also contributed to how each of the three view the other two.

The point remains that it is awkward to discuss the functioning of any of the three rivalries connecting China, the Soviet Union/Russia, and the United States without also considering the state of affairs in the other two.[19] We can insist on examining one rivalry at a time but not without keeping in mind that the rivalries commingle within a major power rivalry field. This view has implications for rivalries among less powerful states in the general Asian region as well.

Other Asian Rivalries

To explore further how this more complicated perspective plays out, we focus on several additional rivalries in this chapter. Following up on the previous section's analysis of the Sino-U.S. and Sino-Soviet rivalries, we turn our attention now to China-Taiwan, China-Vietnam, Thailand-Vietnam, and the two Koreas—all of which shared certain links to the relationships among the United States, China, and the Soviet Union/Russia.

Our intent is to continue examining the applicability of our expectational theory but with the added dimension of rivalry interdependence. We first describe the rivalries briefly and then proceed to assessing the degree of fit of our theoretical expectations. One added question is whether their interdependencies affect our ability to assess the fit of the theory to the observed cases.

The Sino-Vietnamese Rivalry

Like the Sino-Russian conflict, Chinese-Vietnamese hostilities can also be traced some distance back in time, but the current rivalry began a few years after the U.S. withdrawal from Vietnam and the North Vietnamese victory in South Vietnam.[20] During this phase of Indochinese conflict, North Vietnam had received support from both the Soviet Union and China. After its victory in 1975, however, Vietnamese foreign policy increasingly concerned influencing behavior in neighboring Laos and Cambodia. These policies were viewed as a direct threat of Chinese primacy in Southeast Asia. Vietnam and China could not both be the preeminent regional power. This positional issue came to a head when Vietnam invaded Cambodia to eliminate the Khmer Rouge's control in the state.

While China supported the Khmer Rouge, Chinese-Vietnamese relations were further strained by Vietnam's decision to suppress private trading groups in the recently conquered south. The principal target of the new policy were Hoa Chinese in the south. Yet it was Hoa Chinese in the north who began fleeing Vietnam in fear of more repression to come. In response to Vietnam's external and internal policies, China ceased providing aid to its southern neighbor. A few months later, an army of eighty thousand Chinese invaded Vietnam for about half a month, with the expressed intent of teaching Vietnam its proper place in the Southeast Asian hierarchy.[21] The lesson proved costly to China. It also encouraged the forging of an even stronger link between the Soviet Union and Vietnam.

Over the next eight years, Chinese-Vietnamese relations were characterized by shelling across the border, clashes, and a decided reluctance to negotiate on the part of a weaker Vietnam that was increasingly committed to establishing control in Cambodia. The Chinese raised issues of border definitions and the treatment of Hoa Chinese, but the real sticking point was the question of Vietnamese withdrawal from Cambodia. The Chinese were not prepared to negotiate until the Vietnamese withdrew. At the same time, the Vietnamese were not prepared to withdraw prior to the late 1980s and the loss of Soviet patronage and support.

Once Vietnam's military withdrawal was completed in 1989, Chinese-Vietnamese relations returned to some semblance of pre-1977 relations. Boundary questions on land were worked out. At sea, they have remained somewhat more contested with mutual claims over the Spratly Islands in the South China Sea providing a continuing bone of contention. Should the Chinese choose to press their claims to Spratly to the full extent, however, it is not clear that the Vietnamese would be in a position to resist without some form of external support.

The Thai-Vietnamese Rivalry

Thailand and Vietnam had sparred over their relative position in Southeast Asia in general and in the space between them (Cambodia and Laos) for a number of centuries prior to the coming of the period of European control in Southeast Asia.[22] The latest rivalry between the two states dates back at least to the Thai decision to join the U.S. military coalition against North Vietnam. Once that coalition failed to stop the defeat of South Vietnam, the Thai-Vietnamese rivalry began to mirror the Sino-Vietnamese rivalry.

In particular, the withdrawal of the United States in the 1970s made Thai-Chinese cooperation more attractive in the face of expanding Vietnamese activities in states adjacent to Thailand that, historically, had served as something of a buffer between Thailand and Vietnam. As Vietnamese troops engaged Khmer rebels along the Thai border, clashes between Thai and Vietnamese forces became increasingly probable, often on Thai territory. These clashes occurred, occasionally escalated, but never were permitted to become more than border clashes. Once the Vietnamese decided to withdraw from Cambodia, relations between Thailand and Vietnam began to become more normalized and cooperative.

The Chinese-Taiwanese Rivalry

The rivalry between China and Taiwan began when KMT forces were forced to evacuate from the mainland to the offshore island of Taiwan after the civil war that they had lost.[23] Eventually, the new People's Republic of China, established in 1949, might have developed the capability to retake the island, but the Korean War intervened. One of its outcomes was to bestow U.S. protection on Taiwan, now the Republic of China (ROC). The control of the Taiwan Straits by the United States seems to have encouraged a dream of the losers in the civil war that they would return to power on the mainland through the use of military force. But then civil war losers are apt to want to redress the historical outcome in any event. As late as 1962, forces of the Republic of China may have been planning a return to the mainland at a time when the People's Republic of China (PRC) was beset by a number of external and internal problems.[24] During the same period, the PRC initiated intermittent crises (in 1954 and 1958) by shelling small islands off the coast of mainland China but under the military control of the ROC. The crises focused ostensibly on whether the shelling was a prelude to an invasion of Taiwan by the PRC and whether U.S. naval force could deter such an attack. In retrospect, it is not clear that the PRC had the capability to cross the Straits of Taiwan.

Protection of Taiwan by the United States began to be more doubtful after the beginning of the Chinese-U.S. rapprochement in 1972. By the end of the transitional phase of the rapprochement in 1978, the United States had formally severed its official ties to the Republic of China. Yet there is some evidence that the Chinese were prepared to deemphasize the significance of the problem of Taiwan in order to enlist the United States in its competition with the Soviet Union (see Shih 1995: 187).

Gu (1995: 130) characterizes the periodicity of the military conflict in this rivalry as hot war (1949–1969), cold war/standoff (1969–1979), and disengagement after 1980. But the "disengagement" phase has not really persisted, perhaps reflecting competing factions with hard- and soft-line preferences on dealing with Taiwan. The year after Gu published his book, China was sending missiles across Taiwanese air space. The increase in the Chinese-Taiwanese conflict is also partially a function of Taiwanese democratization, which has encouraged some politicians to champion the cause of independence. Intermittently, the PRC reminds the ROC with military maneuvers and missile tests that talk of independence makes the PRC's decision makers unhappy.

Still, as Goldstein (2002) notes, something of an uneasy equilibrium has emerged or reemerged in the case of China and Taiwan in which both sides lack much reason to compromise. At the same time, neither side currently has strong incentives to escalate the conflict, particularly after the Taiwanese elections of 2008 returned the less-independence prone KMT to power in Taiwan. After the election, Chinese-Taiwanese relations definitely improved, but not to the point that China is likely to abandon its insistence on eventual reunification. The question of future conflict then reduces to whether decision makers can successfully manage the equilibrated tensions and/or devise some new formula for resolving the standoff. It is certainly not inconceivable that the standoff could persist for decades even while the two economic systems are becoming more interdependent. There is also the related question of whether the circumstances encouraging non-escalation might change. The U.S. support for Taiwan as long as it does not pursue full-fledged independence has been a critical deterrent that need not always be present or considered credible in the future. Chinese preferences for emphases on economic development and a nonexpansionary public image could be rechanneled into more belligerent stances. A resumption of Taiwanese flirtation with legal independence might also exceed the mainland's tolerance.

The Rivalry between the People's Democratic Republic of Korea and the Republic of Korea

Japan's surrender in World War II led to the presence of Soviet troops in the northern part of the Korean peninsula and U.S. forces in the southern part.[25] Without much apparent premeditation or discussion, this ad hoc division became more concrete when two Korean states were announced in 1948. Both states claimed legitimacy as the sole Korean state. Both states also professed genuine interest in reunification, even though the division institutionalized a more industrialized North characterized by a Marxist regime organized around Kim Il Sung and an agrarian South ostensibly operating within a relatively democratic regime that has since oscillated between civilian and military rule.

In 1950, the better-armed North attempted a coercive reunification of the two states. Three years of highly destructive warfare ensued that brought in military intervention by the United States and China, among other states. The armistice of 1953 merely solidified the cleavage between the two Koreas. It also forced both sides to devote considerable resources to repairing the wartime devastation for the next twenty years. Initially, the more industrialized Northern economy was more successful in realizing growth, but it developed serious problems in the 1970s while at the same time it lost some of its external support and suffered from a string of environmental catastrophes. Southern industrialization and economic growth was initially slower but ultimately more spectacular and certainly successful in establishing a modern, competitive economy.

Throughout these economic gains and vicissitudes, the bilateral relations between the two Koreas have alternated between episodes of relative harmony and negotiations toward reunification and intense hostility. On at least three occasions, increasingly serious efforts have been initiated to bring about a de-escalation of tensions between the North and the South. Prior to 1971, the North had insisted that the South's ruling elite would have to lose power before any discussions could hope to succeed (Oberdorfer 2001). This insistence was relaxed in the first attempted rapprochement in 1971–1972. The second attempt began in 1988 on the South's initiative, but it did not gain much momentum prior to 1991. By 1993, this second interlude seemed to have run its course. A summit meeting involving North and South Korean leaders in 2000 initiated a third attempt that eventually was caught up (and bogged down) in maneuvers by major powers concerning North Korea's nuclear threat (Heo and Roehrig 2010: 129–156). Newly elected South Korean leaders moved toward more conservative and less accommodating positions vis-à-vis North Korea. In turn, North Korean attacks, verbal and physical, became more heated, including the sinking of a South Korean naval ship (2009) and the shelling of a populated island (2010). A change in leadership in North Korea at the end of 2011 offered an opportunity for changed behavior but, to date, the North's behavior towards the South does not appear to have been altered in any significant way.[26]

Yet the history of North and South Korea is also characterized by an unusually long series of Northern raids and incursions after 1953 that have always stopped short of a resumption of warfare.[27] At the same time, both sides have maintained large military forces and prepared for a renewed outbreak of intense conflict—mitigated presumably by the presence of a large number of U.S. troops that have been stationed in the South to serve as a deterrent to another Northern invasion.

In the past decade or so, international attention has focused on North Korean attempts to build nuclear weapons that could threaten Japan and even the United States. Concurrently, the Northern economy has continued to deteriorate and political control was passed from father to son (from Kim Il Sung to Kim Jong Il and then to Kim Jong Un). Even so, the two Koreas continue to contemplate at least the rhetoric of reunification while preparing for the possibility of renewed warfare either of an intra-Korean nature or, possibly, involving a preemptive strike by the United States on the North.

External Factors

All of these rivalries can be described in terms that are restricted to the specific rivalry in question. Yet there are or have been significant external factors in their functioning. The very existence of a Chinese-Taiwanese rivalry is predicated on U.S. support for Taiwan. The extent of U.S. support for Taiwan has not been constant; it appears to be correlated strongly with Sino-U.S. hostility. Taiwan became more important to the United States after the Chinese intervention in Korea. When the United States sought to de-escalate its rivalry with China, U.S. support for Taiwan was awkward and played down. The renewal of Sino-U.S. rivalry in the second half of the 1990s is linked in part to the continued U.S. defense of Taiwan.

Vietnamese ambitions and operations in Southeast Asia after the end of the Vietnam War were undergirded by a strong relationship with the Soviet Union. The weakening and ultimate collapse of the Soviet Union meant that Vietnam was forced to reconsider its defiance of Chinese preferences. It is hardly coincidental that China and Vietnam de-escalated their rivalry within a context of Soviet decline.

The Thai-Vietnamese conflict had clear antecedents in history, but the latest manifestation was closely geared to U.S. support for Thailand during the Indochinese warfare of the Cold War era. The decline of U.S. support was not sufficient to bring about de-escalation, but the Vietnamese loss of Soviet support contributed to Vietnam's pursuit of a less coercive bid for leadership in the region. In turn, this diminished the level of threat perceived in Thailand and created a situation conducive to rivalry termination.

North Korea's earlier ambitions and more recent intransigence have been predicated on support initially from both China and the Soviet Union. The decline of the Soviet Union meant that North Korea could no longer expect the two Communist states to compete in providing external assistance. North Korean decision makers subsequently became more open to South Korean overtures. Similarly to the Taiwanese situation, South Korea would have been conquered by North Korea if it had not been for U.S. intervention and continuing military presence. As a consequence, relations between South Korea and North Korea are caught up in a triangle linking the two Korean states with the United States that often also resembles a quadrangle connecting to China as the fourth state. North Korea sometimes threatens South Korea to influence South Korean behavior, but it also uses South Korean vulnerability as a threat to obtain attention and concessions from the United States and perhaps also China. Problems in the Korean peninsula lead the United States and China to attempts to encourage each other to either restrain North Korea (United States to China) or to desist from threatening North Korea (China to the United States). It would thus be very difficult to interpret the functioning of the rivalry of the two Koreas solely as an inter-Korean affair.

We stress these complications of regional context because it is difficult to view de-escalation processes as taking place strictly within any given rivalry. Rivalries are variably interdependent throughout the world. In Asia, they tend to be closely linked to the rivalries among the major powers because the major power rivalries are themselves highly interdependent and because the major power patrons have clients that are also engaged in rivalries. None of these complications preclude, however, the application of an emphasis on shocks, expectancy revision, reciprocity, and reinforcement as the key to explaining de-escalation/termination. That is, the contextual complications do not preclude the application of the model as long as we keep in mind that some rivalries are nested within other rivalries. We should expect to link shocks and expectancy revisions to changes ongoing in other rivalries in the same neighborhood.

Analysis

In any event, the four additional rivalries (China-Taiwan, China-Vietnam, Thailand-Vietnam, and North Korea–South Korea) yield six more cases relevant to rivalry termination questions. Two terminations, three temporary de-escalations, and one no change provides some welcome and, presumably, useful variation in outcome. Our primary analytical question is whether, or to what extent, these additional six cases are associated with shocks, expectancy revision, consolidated change-seeking entrepreneurs, third-party pressures, reciprocity, and reinforcement. A second question is whether this list omits anything of possible theoretical significance.

Shocks are abrupt changes in internal or external environments. They are capable, as we have seen in earlier chapters, of stimulating decision makers to rethink their strategies and perhaps to change their strategies. Yet, obviously, there is no guarantee that decision makers will respond to these shocks in any meaningful way. Nor is there any expectation that they will always respond either positively or negatively. We cannot even be sure that they will respond uniformly, predictably, or at all. The real question, however, is whether decision makers are likely to have incentives to overcome their policy inertia in the absence of some abrupt stimuli that provokes reevaluations of ongoing approaches to dealing with external enemies (and friends).

In this respect, eastern Eurasia of the past few decades is rich with various types of shocks. The most persistent type of shock has been caused by rapprochements among the United States, China, and the Soviet Union.

Each time this has occurred, client states have to wonder whether their client status has been jeopardized by increased collusion among formerly hostile patrons. Could they continue to compete with their rival if external support and protection vanished? The movement of the United States and China toward rivalry de-escalation beginning in 1971 constituted a shock for the Chinese-Taiwanese rivalry and the first Korean case. It also facilitated the subsequent emergence of the Sino-Vietnamese rivalry. Taiwan was threatened with a loss of U.S. protection. The two Koreas were unsure how to forecast the implications for themselves. The worst-case scenario was that both North and South might lose support from China and the United States, respectively.

The second shock of this nature was generated by the Sino-Soviet rapprochement of 1989. This one influenced the Korean, Chinese-Vietnamese, and Thai-Vietnamese rivalries. North Korea had managed to utilize the Sino-Soviet rift to its own advantage, securing resources from both sides, and maintaining some autonomy as the two Communist giants competed to support North Korea. The disappearance of that rivalry and the desire by both China and Russia to prioritize domestic economic development over foreign commitments spelled increasingly hard times for North Korea.

In the wake of the end of the Sino-Soviet rivalry, the Soviet Union had begun to improve its relations with South Korea substantially. Soviet trade with North Korea decreased dramatically. Peaking in 1989, Soviet–North Korean trade declined by half in 1990, and then to a third of the 1990 figure in 1991. Since the Soviet Union had been the primary source of oil, coal, electric power, steel, and related commodities, North Korean industry could hardly go unaffected even if some of the deficit was made up by Chinese imports. North Korean gross domestic product declined each year through most of the 1990s as a partial consequence (Zacek 1998: 82–84). Soviet/Russian military and economic assistance diminished abruptly. Chinese aid to North Korea also declined. As Eberstadt (1998: 245) puts it: "The sudden and dramatic sequence of world-wide events between 1989 and 1991 associated with what is now called 'the end of the Cold War' represented for Pyongyang, a catastrophe of previously unimaginable scope and proportion" (see also C. Armstrong 2004: 45).

The impact on the Chinese-Vietnamese rivalry was somewhat similar in magnitude if not in kind. Vietnam could challenge China's much greater resource base only with the support of the Soviet Union. Once that patronage was reduced drastically, Vietnam had seemingly few options but to work out some form of accommodation with its very large neighbor to the north. Doing that required a withdrawal from Cambodia/Kampuchea. In turn, this retreat from Cambodia changed the nature and likelihood of Thai-Vietnamese conflict. Vietnamese troops would no longer be likely to make incursions into Thai territory in pursuit of Khmer Rouge rebels. A dampening of Vietnamese ambitions of regional preeminence in Southeast Asia also appeased both Chinese and Thai perceptions of threat. Wattanayagorn (1998), for instance, notes that Thai elites no longer regarded the eastern threat emanating from Vietnam after the late 1980s as a serious one.

Economic resource crises appear to have facilitated shifts of varying degrees in the priorities of regimes in five cases. Crop failure and famine in the 1990s certainly increased North Korean vulnerability. Domestic Vietnamese economic problems emerged prior to the loss of Soviet patronage and trade concessions that only made the economic problems worse.[28]

Shifts in regime orientations could also occur without the stimulation of economic deterioration. Taiwan became decidedly more democratic in the mid- to late 1990s.[29] Yet asymmetrical democratization seems only to have exacerbated the Chinese-Taiwanese rivalry. Taiwanese politicians now have an increased structural incentive to bid, and perhaps to outbid one another, for championing the case for independence. The rivalry has shifted into a phase in which hostilities can be suppressed as long as Taiwan maintains a low profile. Assertions of independence antagonize the mainland Chinese and lead to some type of warning and/or demonstration of China's coercive potential. Not all regime shifts, therefore, are necessarily helpful for rivalry termination purposes.[30]

In both the Chinese-Vietnamese and Thai-Vietnamese cases, changes in regime personnel appeared to have strengthened the case for one side accepting the other side's shift in priorities as genuine. In Vietnam, generational shifts in elites helped remove individuals who were opposed to cooperating with the Chinese. In Thailand, a new civilian leader was perceived to be less enamored of military strategies than earlier military leaders (Porter 1993: 214).

Table 7.5 thus summarizes the six-case outcome for the question of shocks. All the cases experienced shocks of some kind. Two clusters of cases emerge readily if the six cases are arrayed from no change to termination (as they are in Table 7.5). Four cases (China-Taiwan and the three Korean cases) experienced one or two shocks, but the cases with two shocks did not necessarily come at the same time. Two cases experienced all four of

Table 7.5. Shocks in the Eastern Asian Rivalries

Rivalry	*Shift in external threat*	*Shift in competitive ability*	*Shift in regime orientations/ strategies*	*Economic resource crisis*
China-Taiwan		Present	Present (Taiwan)	
N. Korea–S. Korea 1		Present		
N. Korea–S. Korea 2		Present		
N. Korea–S. Korea 3		Present		Present (N. Korea)
China-Vietnam	Present	Loss of Soviet support (Vietnam)	Present	Present (Vietnam)
Thailand-Vietnam	Present	Loss of Soviet support (Vietnam)	Present	Present

the highlighted shocks. The latter cluster is associated with significant deescalation or termination; the former cluster is associated with limited or no change. Other things being equal, then, Table 7.5 suggests that compounded or multiple shocks can be quite important to altering rivalry endgames. Singular or isolated shocks, other things being equal, are more likely to be weathered or at least insufficient in motivating decision makers to seriously contemplate doing things differently.

The outcome for observed expectancy revision resembles the distribution of compounded shocks, which were associated with the cases in which some definite change was registered. There has been only limited expectancy revision in the Chinese-Taiwanese case or the first two Korean cases. In both rivalries, the stakes imply one or both sides ultimately surrendering its autonomy to a larger entity. While decision makers on Taiwan presumably no longer dream of reconquering the mainland, they still see the PRC as Taiwan's primary external threat, and certainly not without reason. The PRC has experimented with different approaches to dealing with the Taiwanese issue but has never retreated from its basic position that Taiwan is a part of greater China.[31]

The Korean rivalry faces a similar dilemma with the difference that two separate states is a more plausible future possibility. In the early 1970s, both sides were willing to negotiate for fairly limited, short-term objectives that amounted to threat reduction. Both sides appreciated the value of relaxed tensions for external audiences and, in the case of South Korea, internal audiences as well. But there does not appear to have been any persistent interest (as opposed to temporary attempts) in making major concessions to the other side on either party's part. In the early 1990s, more headway was made on negotiations thanks considerably, no doubt, to North Korea's sense of heightened vulnerability. But the cooperative interactions could not be maintained in the mixed context of international pressures pertaining to North Korea's nuclear weapons program and domestic political pressures within South Korea. In terms of the nuclear weapons, North Korea was being accused of not completing its end of the bargain and, in return, aid to the North was being suspended. North Korea's response was to escalate the verbal tenor of its nuclear threat to South Korea. In South Korea, Roh Tae Woo was in the process of being replaced by Kim Young Sam, who thought that improved relations with North Korea would benefit his domestic rival, Kim Dae Jung. The South Korean military was also pressing for a resumption of "Team Spirit" exercises—combat simulations by the United States and South Korea that were perceived in the North as practicing for an invasion of North Korea. The North Korean response was to break off further intra-Korean negotiations.[32]

The third Korean episode of attempted de-escalation was led by Kim Dae Jung (Snyder 2004: 24) after his coming to power in 1998. A summit meeting between Kim Dae Jung and Kim Jong Il in 2000 initiated a "Sunshine Policy" period of substantially reduced threat perception at least on the part of the South Korean public and increased trade and tourism (Cha 2003: 114). Yet Chung-in Moon and Taehwan Kim (2003: 256) write that Kim Dae Jung's efforts did not lead to "any fundamental changes in threat perception, force structure, deployment patterns, and military planning in the two Koreas." Nor did the sunshine orientation long survive Kim's term of office. In the second half of the twenty-first century's first decade, inter-Korean relations have deteriorated considerably and returned by 2010 to mutual threats of going to war.

In contrast, China, Vietnam, Thailand, and the Soviet Union/Russia all saw the levels of threat emanating from their respective opponents as diminishing. Vietnam ultimately acquiesced more or less to Chinese preferences in Cambodia and, therefore, had much less reason to expect a renewed invasion along the lines of 1979 or the intermittent border shelling and clashes that followed that punitive expedition. Chinese decision makers were prepared to accept at least partial Vietnamese subordination.[33] The Vietnamese threat to Thai interests visibly receded with the Vietnamese withdrawal from Cambodia.

In this group of rivalries, the presence of change-seeking political entrepreneurs was most visible in the Korean efforts of Kim Dae Jung to play a role with some similarity to Gorbachev's, but, ultimately, he failed to overcome the various obstacles to Korean de-escalation of tensions.[34] Some hard-liners were replaced in the Vietnamese and Thai political systems at times that facilitated more cooperative stances with external adversaries, but these situations seem to fall short of what is required for consolidated change-seeking leaders.

Third-party pressures do not appear to have played an extensive role in terms of actively encouraging de-escalation between rivals, with the obvious exception of the Chinese-Vietnamese case in which the Soviets desired the Vietnamese withdrawal as part of the price of their own bargaining with the Chinese. Scobell (2004: 83–84) also credits Chinese encouragement for facilitating the Korean summitry of 2000 and argues that the North Koreans would not have been receptive to Kim Dae Jung's initiative without considerable Chinese support.

This is not to say that third-party pressures have been completely absent, but these efforts have not been well aligned with local preferences for reductions in hostility. Perhaps more important is third-party neutrality or lack of serious opposition to de-escalation talks. This factor is more difficult to evaluate, but in situations in which patron-client relationships have loomed large, such as the rivalries involving Taiwan or the Koreas, it stands to reason that de-escalation maneuvers would need at least the absence of significant opposition by patrons to make much headway. In this spirit, Oberdorfer (2001: 260) writes that the winter of 1991 was "one of the rare periods when the policies of the two Koreas were in alignment for conciliation and agreement, with all of the major outside powers either neutral or supportive." Not coincidentally, the same evaluation could not be made of the major power environment applicable to the third Korean case.[35]

Reciprocity and reinforcement require that some initial progress be made in de-escalating rivalry relationships. These aspects, therefore, do not apply to the Chinese-Taiwanese case, which has yet to proceed that far. In the three ultimately aborted Korean efforts, each time there were major restrictions on the extent to which one or both sides were prepared to go, despite an outright attempt to purchase cooperation in the third case. All three efforts had limited life spans. Reinforcement was at best of temporary duration.

Again, the more successfully de-escalated cases cluster positively on the reciprocity and reinforcement dimensions. All three cases have forged ahead, settling lesser disputes and backing away from possible intensive disagreements.[36] As a consequence, no relapse in their relationships have occurred even though it is not clear that one of the two cases involving Vietnam is fully terminated.

Table 7.6 summarizes the six factors that the expectancy theory highlights as most significant and their presence or absence in the six Eastern Asian cases. The overall pattern that emerges is quite strong. Shocks are found everywhere, but multiple, compounded shocks are found only in the Thailand-Vietnam and China-Vietnam rivalries. The same locational concentration applies to expectancy revision. Consolidated, change-seeking entrepreneurs and third-party pressure are found here and there, but no consistent pattern emerges other than that they can be facilitative, especially in the case of the policy entrepreneurs. The presence of reciprocity and reinforcement, in contrast, are concentrated in the lower right-hand section of the table. In sum, the outcomes associated with these six cases suggest that compounded shocks, expectancy revision, reciprocity, and reinforcement are necessary, but there is no evidence that any are sufficient causes of rivalry de-escalation and termination. All of these outcomes conform to theoretical expectations—thereby providing support for the utility of the expectancy theory.

Does anything significant appear to be missing from the theory? The compounded nature of shocks was not predicted specifically. The theory states that major shocks may be needed to tip expectancy alteration processes. There is certainly no incompatibility between "major" and "compounded" shocks. Compounded shocks merely help to delineate more precisely what sort of major shocks are most effective, as well as how "major" they need to be. The shocks also need to be across the board, as opposed to a single type no matter how acute. It apparently helps to overwhelm expectancy rigidities with substantial and multidimensional environmental change.

Yet even overwhelming change is not sufficient. The case of North Korea, and to a lesser extent, that of Taiwan, are evidence of that. In the 1990s North Korea confronted losses of patronage and imports, environmental disaster, and increased external hostility over its nuclear policies. In some

Table 7.6. Modeling Outcomes for Other Asian Rivalries

Variable	*China-Taiwan*	*N. Korea–S. Korea 1*	*N. Korea–S. Korea 2*	*N. Korea–S. Korea 3*	*China-Vietnam*	*Thailand-Vietnam*
Shocks	Present	Present	Present	Absent	Present	Present
Expectancy revision	Absent	Absent	Some	Present	Present	Present
Consolidated and new policy entrepreneurs	Absent	Absent	Limited	Present	Absent	Absent
Third-party pressure	Absent	Absent	Absent	Present (China)	Absent	Absent
Reciprocity	Absent	Some	Some	Present	Present	Present
Reinforcement	Absent	Absent	Absent	Presentfor a few years	Present	Present
Outcome	No change	Limited change	Limited change	Temporary change	Termination	Termination

respects, its situation was comparable to Vietnam's in the same decade. But Vietnam chose to de-escalate while North Korea remained more intransigent, albeit more with major powers than with its rival to the south. Dissimilarly, Vietnam's significant de-escalations with China and Thailand have remained operative while the Korean sunshine interlude fell on hard times within a few years. Taiwan has also experienced major problems with the circumstances of its external patronage without yielding to the huge asymmetries associated with the Chinese-Taiwanese rivalry.

One factor, then, that is present in the North Korean and Taiwanese cases but was absent in the Sino-Vietnamese and Thai-Vietnamese cases is a matter of regime survival. The Chinese did not demand that the Soviets or the Vietnamese alter the nature of their political-economic regimes. On the contrary, in both cases, the Chinese strongly preferred that their adversaries remain very much in the socialist camp.[37] Neither the Thai nor the Vietnamese governments expected their adversaries to alter radically the nature of their regimes. The Vietnamese actually welcomed the ascent of civilian party leaders with strong interests in expanding trade opportunities, figuring them to be easier to deal with than earlier military leaders. The point here is that decision makers could contemplate rivalry de-escalation without also being forced to accept political seppuku.

The Taiwanese decision makers have not yet really been offered a choice between continuing rivalry and de-escalation without some form of subordination to the Chinese mainland. Initially, at least from the mainland's perspective, it was reincorporation or else, which then evolved into the possibility of a "Hong Kong" arrangement, or reincorporation with some autonomy. Yet autonomy without reincorporation remains a more attractive option for the Taiwanese as long as it can be maintained. This status may be sustainable as long as Taiwan does not forcefully declare its independence from China.[38]

The North Korean case involves a set of elites who have been unsuccessful in the economic development of their country despite inheriting more than their share of Korea's industrial potential, and yet they continue, for the most part, to pursue strategies that promote economic autonomy. Presumably, and partially as a consequence, they depend on external threats and military force to remain in power. Their rival to the south was once impoverished and agrarian. Now it is an industrial power that is anything but impoverished. Genuine rapprochement with the South, therefore, could easily lead to the downfall of the northern political regime. The Northern decision makers' dilemma, then, is how to manage their multiple problems—internal and external—without voluntarily surrendering the northern regime's very existence.[39]

The North Korean case also suggests a second factor. Vietnam capitulated its foreign policy ambitions (but not its regime) because few options were available once Soviet patronage disappeared. Vietnam had failed to make any alternative arrangements with the United States at that point. Its choices were to yield or go down fighting an array of internal and external problems. In the early 1990s, North Korea might have been seen as facing a similar type of problem. It, too, had lost critical Soviet patronage. But it had two other options that were not available to Vietnam. Chinese patronage continued although not at previous levels (Kim 2004: 204). There was also the nuclear gambit that could be explored by a regime that was quite accustomed to what has been termed a mendicant foreign policy. The survival of the regime prior to the 1990s had depended on Soviet and Chinese aid. Perhaps after the 1990s it could depend on Chinese, western, and South Korean aid.[40] In other words, there were perceived alternatives that could be explored experimentally and that suggested outcomes that involved, with some luck, neither yielding nor going down fighting.

We might refer to these additional factors as regime survival and geopolitical idiosyncrasies. The former appears to be a rather potent barrier to expectancy revision. Rivalry de-escalation and/or termination appears to be least probable in situations that require one or both sides to surrender voluntarily not only the issues in contention but also their political system's regime identity and decision makers' own positions of entrenched power. The geopolitical factor is more situational in nature and thus less amenable to generalization, except to say that some local geopolitical circumstances can shield decision makers from the full implications of shocks—or at least give them reasons to believe that they are partially insulated from these implications if multiple great powers are present and heavily involved in the foreign affairs of weaker client or allied states. When de-escalation becomes more a matter of triangular (or multiple triangular) interactions than dyadic ones, the situation may well call for a different kind of theoretical focus than the one implicit to the expectations theory.[41]

Do we need to make these amendments to the expectancy theory? They do not appear to be critical at this juncture. Should they appear as well in other rivalry environs, however, it would not be difficult to add them as refinements to the basic argument that portrays shocks and expectations as most critical to rivalry de-escalation, along with reciprocity and reinforcement once the process is underway. So far, though, the basic argument has held up reasonably well. Some factors appear to be necessary while others are either facilitative or get in the way of successful de-escalation. Only more examinations of more rivalry cases can confirm the ultimate utility of the theory, but we still appear to be on the right track.

CHAPTER 8

The Outcome: Assessing the Rivalry De-escalation Theory

In Chapters 1 and 2, we presented and developed our argument that some types of interstate rivalry terminations can be explained with a small handful of variables. It was noted that a respectable number of rivalries terminate simply because one or both sides acquiesce to the superior strength of their adversary. They are beaten in wars, eclipsed by transition dynamics, or simply are forced to acknowledge that they are unable to maintain their competition. In other words, they are no longer in the same league. These rivalry terminations are not difficult to explain. Other rivalries, however, cannot be explained by coercion, subordination, or surrender. This other class of rivalries is subject to negotiations and decisions to de-escalate. Our model is designed to encompass this other, less coercive category.[1]

The model focuses on the revision of one or both sides' expectations about the other party. Since interstate rivalry is predicated on the perception of competitors that are also threatening enemies, the revisions tend to focus on downgrading the perceived degree of threat or enemy status. Changes in expectations do not necessarily mean that one's rival is suddenly embraced as a friend or ally. Negative images are likely to persist. But, the discernment of decreased threat or enemy-like behavior, in conjunction with a small set of additional variables, can lead to the termination or significant de-escalation of a rivalry.

Expectational change, while essential to our argument, is not sufficient. We think some mixture of the following is needed to bring about voluntary de-escalations. In addition to the expectancy revisions, shocks are needed to overcome the pronounced tendency toward foreign policy inertia. The more ingrained the inertia, the greater the shock or shocks that need to take place. Policy entrepreneurs can certainly help organize support and lead in overcoming previous stances taken toward a one-time enemy. Third-party intervention may also help this process of change gain and maintain momentum, perhaps through the provision of resources and protection—or the threat of withholding one or the other commodity. Yet neither policy entrepreneurs nor third-party intervention are essential. They can facilitate the policy change, but they do not appear to be absolutely necessary. More critical to the outcome, we think, is the need for an adversary to reciprocate any concessions offered and to continue reinforcing the other state's decision to move in a different policy direction. In the absence of reciprocation, little in the way of progress can be made in revising how states treat each other. Once some progress is made, de-escalating rivals need to continue avoiding behavior that will alarm their adversaries and give grounds for setting back the movement toward de-escalation.

Our argument is reasonably parsimonious and, more importantly, testable.[2] How does it fare when the argument is confronted with some thirty-three Eurasian cases with different outcomes (within ten rivalries) that have been described in Chapters 3 through 7? There are definitely ways to analyze thirty-three cases in the aggregate that are more sophisticated. However, we find the outcome to be unusually clear in its pattern—so clear that sophisticated methods hardly seem warranted. Instead, we break down the outcome into three groups and tables. The first table (8.1) examines only the twenty-two cases in which the rivalry experienced no change. Two additional tables focus exclusively on eleven cases in which some de-escalation was achieved (seven cases in Table 8.2) or termination was accomplished (four cases in Table 8.3).

Table 8.1 summarizes the outcome associated with cases in which most of the relevant explanatory variables should be recorded as absent. In all but one column, the collective values of the variables are 80 percent or more absent. The nonconforming column pertains to shocks, which are quite common in Table 8.1 (86 percent). By and large, then, it should be clear that shocks are simply opportunities at best. It depends on whether anyone is willing or eager to exploit them. In Table 8.1, the "no change" cases were situations in which some effort was made to consider de-escalation. In most cases, only the shock column registers a positive value. With but two exceptions (Egypt-Israel, Sadat 1; and Israel Syria, 1992–1996), the other positive values are scattered in situations in which only one or two of the six explanatory variables are coded as positive. The Sadat 1 exception is an asymmetrical situation. Sadat was prepared to negotiate in the early 1970s but his Israeli counterparts were not. The "solution" was to create a new shock in the form of the war of 1973, which brought with it the additional bonus of third-party diplomatic intervention. The second exception lacked a shock that might disrupt foreign policy inertia. None of the other values, and especially the strong mistrust associated with both sides' expectations, was strong enough to bring about de-escalation in the absence of a shock or shocks. Table 8.1 thus passes our first test quite well. The rivalries that did not de-escalate significantly have been predicted as not changing much by our model.

Table 8.2 isolates the cases in which some de-escalation was experienced. In this situation, we should expect a fairly high proportion of positive scores for the explanatory variables. Since the outcome is "some or limited" de-escalation, there is no reason to anticipate the variables to be present at full strength. Something should be missing.

One possible pattern is a block of cases with some present "hits" on some but not all variables. Another possible pattern involves some cases with a strong array of predictors being present and some with a mixed but less than full array. It is the latter that is demonstrated in Table 8.2. The first three cases listed in the table show the presence of all six variables and some de-escalation. The other four instances include one Indo-Pakistani episode and three Korean cases. All seven instances represent hard cases of strongly entrenched rivalries. In each case, there was substantial resistance to full termination, albeit variably manifested especially in the Korean cases. In the Egyptian-Israeli case, the movement toward de-escalation went only so far. A "hot war" environment moved to a "cold peace" era without fully ending the rivalry. Momentum toward de-escalation in the Indo-Pakistani case has never proceeded very far. In the case of the two Koreas, substantial progress was made toward de-escalation after a couple of false starts until a change in regime in South Korea elevated to power a more conservative group that had always been less inclined to reinforce the process of de-escalation. In that respect, the Korean case resembles the Israeli-Palestinian one.

What seems most important about these cases is that there were limits to the degree of expectancy revision achieved. The Egyptians and Israelis still regard each other as strategic rivals. What has changed is that a combination of U.S. intervention and substantial foreign aid, Israeli security preferences, and Egyptian development preferences came together to foster an

Table 8.1. The Summary Empirical Outcome: Rivalries Involving No Change

Rivalry	*Shocks*	*Expectancy revision*	*Consolidated and new policy entrepreneur*	*Third-party pressure*	*Reciprocity*	*Reinforcement*	*Outcome*
Egypt-Israel Nasser	Yes	No	No	No	No	No	No change
Egypt-Israel Ben-Gurion	Yes	No	No	No	No	No	No change
Egypt-Israel Sharett	Yes	No	No	No	No	No	No change
Egypt-Israel Eshkol	Yes	No	No	No	No	No	No change
Egypt-Israel Meir 1	Yes	No	No	No	No	No	No change
Egypt-Israel Sadat 1	Yes	Yes	Yes	No	Yes	Yes	No change
Egypt-Israel Meir 2	Yes	No	No	No	No	No	No change
Israel-Syria 1949	Yes	No	No	No	No	No	No change
Israel-Syria 1952	Some	No	No	No	No	No	No change
Israel-Syria 1967	Yes	No	No	No	No	No	No change
Israel-Syria 1974	Yes	No	No	Some	No	No	No change
Israel-Syria 1990–1991	Yes	No	No	Yes	No	No	No change
Israel-Syria 1992–1996	No	Some	Some	Yes	Some	No	No change
Israel-Palestine	Yes	No	Yes	Yes	Some	No	No change
India-Pakistan 1962–1963	Yes	No	No	Yes	No	No	No change
India-Pakistan 1966	Yes	No	No	Yes	Some	No	No change
India-Pakistan 1980	Yes	No	No	No	No	No	No change
India-Pakistan 1989	Yes	No	No	No	No	No	No change
India-Pakistan 1999	Yes	No	No	No	No	No	No change
India-Pakistan 2003–2007	No	No	Yes	No	No	No	No change
India-Pakistan 2010	No	No	Yes	No	No	No	No change
China-Taiwan	Yes	No	No	No	No	No	No change

Table 8.2. The Summary Empirical Outcome: Rivalries Involving Some De-escalation

Rivalry	*Shocks*	*Expectancy revision*	*Consolidated and new policy entrepreneur*	*Third-party pressure*	*Reciprocity*	*Reinforcement*	*Outcome*
Egypt-Israel Sadat 2	Yes	Yes	Yes	Yes	Yes	Yes	Some de-escalation
Egypt-Israel Rabin	Yes	Yes	Yes	Yes	Yes	Yes	Some de-escalation
Egypt-Israel Begin	Yes	Yes	Yes	Yes	Yes	Yes	Some de-escalation
India-Pakistan 1972	Yes	Some	No	No	Limited	Limited	Some de-escalation
N. Korea–S. Korea 1	Yes	No	No	No	Some	Some	Limited de-escalation
N. Korea–S. Korea 2	Yes	Some	No	No	Some	Some	Limited de-escalation
N. Korea–S. Korea 3	No	Yes	Yes	Some	Yes	Halted	Temporary de-escalation

agreement not to threaten each other with war once Egypt regained its lost territory. As long as Egypt remains outside of a possible attacking coalition, Israeli security is improved because a conventional Arab military attack is unlikely without Egyptian participation. Thus, the Egyptian-Israeli de-escalation is a very special case.

The Indians and Pakistanis, so far, have not had much reason to change their respective evaluations of their adversary's intentions or threat potential. The nearly decisive defeat of Pakistan in 1972 might have led to a situation in which rivalry termination was related to a defeat in war—the category our model is not trying to explain. But the basic issues at stake, especially Kashmir, did not go away and were revived by independent events in Kashmir and not in Pakistan. At the same time, the perceptions of the two rivals of their adversary have never altered all that much. Once the Kashmir revolt was renewed by Kashmiris, the Indo-Pakistani rivalry was back on at full strength again.

The Korean circumstances are different. The first attempt at de-escalation was not accompanied by any expectancy revision. The next one involved some (at least in the south) and the third was predicated on considerable change on the part of the South Korean government. The problem was that there was no guarantee that the attitude of the South Korean government would remain the same after several subsequent elections. Another way of putting it is that progressive/liberal expectancies about North Korea changed substantially while conservative expectancies changed less. No one seems too sure of what North Korean expectancies were or to what extent they changed. But as long as progressive/liberal governments remained in power in South Korea, rivalry de-escalation might proceed. Once a conservative regime gained power in the south, the rivalry returned to its old format of intermittent North Korean attacks on South Korea.

Four cases of full termination are listed in Table 8.3. In this test we should expect a high proportion of positive values. The predicted four of the six explanatory variables are associated with "perfect" outcomes in terms of our four core variables. Shocks, expectancy revisions, reciprocity, and reinforcement are all present in the termination cases. Presumably, they are all necessary, as we have predicted, although none alone is likely to be sufficient. Policy entrepreneurs and third-party intervention, however, are sometimes present and sometimes not. We had thought initially that policy entrepreneurs and third-party intervention could be extremely helpful but that neither was absolutely necessary. That expectation seems to be supported in

Table 8.3. The Summary Empirical Outcome: Rivalries Involving Termination

Rivalry	*Shocks*	*Expectancy revision*	*Consolidated and new policy entrepreneur*	*Third-party pressure*	*Reciprocity*	*Reinforcement*	*Outcome*
China–United States	Yes	Yes	Yes	No	Yes	Yes	Termination
China–Soviet Union	Yes	Yes	Yes	No	Yes	Yes	Termination
China-Vietnam	Yes	Yes	No	Some	Yes	Yes	Termination
Thailand-Vietnam	Yes	Yes	No	No	Yes	Yes	Termination

Table 8.3. In some cases, these two variables may have been essential.[3] But, in general, they do not seem to be as crucial as the hard core of the four critical variables.

What these findings add up in terms of our model is rather strong support for the combined presence of shocks, expectancy revision, reciprocity, and reinforcement. Other factors, such as policy entrepreneurs and third-party intervention, are less critical to the outcome. Still, our model is meant to be highly parsimonious. It does not, for instance, attempt to capture regional landscapes (Chapter 7) per se that may also work toward or away from rivalry de-escalation. The comings and goings of major power rivalries or, for that matter, major powers, tend to have reverberations for rivalries involving minor powers. These "earthquakes," of course, can appear in the parsimonious model as shocks for rivalry relationships. But the basic point to be stressed here is that rivalries are not maintained in dyadic vacuums in which all the important factors that drive rivalries are endogenous to the protracted conflicts.

We have few studies that examine systematically the transition of protracted conflicts from escalation to de-escalation. Of those that have, none have examined the transition more thoroughly than Kriesberg's studies (1998, 2007). Not unlike the approach herein, Kriesberg (1998) argues that de-escalation occurs as a result of changes in the conditions that underlay the emergence of a conflict in the first place and sustained its escalation. He goes on to say, however, that de-escalating transformations are long, cumulative processes. They are not brief, clearly delineated events. The shift toward de-escalation is produced by pressures building over time. If there are astounding moments of change, then they usually follow from many less visible trends (Kriesberg 1998: 190). In short, the de-escalation of intractable conflicts is not necessarily the result of immutable, large-scale forces or of the actions of a few brave individuals. Many circumstances need to converge and these must be interpreted in new ways in order for intractable conflicts to move toward de-escalation (Kriesberg 1998: 216–217). Only in retrospection can we discern a long-term transformation toward de-escalation (Kriesberg 1998: 24).

Our model would not disagree with many of the points made by Kriesberg (1998). While we would agree that changes in macro-conditions must converge with new interpretations, we do not necessarily support the notion that transformations always entail long-term processes or that their effects have to be cumulative over time. On the contrary, change can occur quickly and be observable almost immediately. This is especially the case when environmental crises affect the political survival of policymakers. In these instances, leaders have strong incentives to change their strategies and policies quickly. Although the shifts in their strategies or policies may not generate immediate outcomes of conflict termination, they can move protracted conflicts in new directions in relatively short periods of time. Our perspective leaves open the possibilities for short- and long-term effects because the analytical effort is not to explain the precise moment at which conflicts end but rather the process by which protracted conflicts eventually end.

However, an expectancy model does not have to be so imprecise that it cannot pinpoint moments of change. Political shocks tell us where the important turning points will or may be; they delineate the moment at which decision makers are more likely than at any other time to reevaluate their expectations, strategies, and policies. Without the environmental challenges that shocks pose, inertia overrides the inclination of decision makers to pursue new, risky policies that could undermine their positions. Therefore, expectancy models are indeed historically contingent, but they are not so idiosyncratic that we cannot specify generalized patterns of behavior across many varied cases of protracted conflict. Moreover, the interactions between shocks and expectations are hardly the end of the de-escalation process. Rather, it is merely the beginning. The question then concerns what actors and actions are critical to keeping the de-escalation process in motion.

It is always gratifying to learn that a parsimonious model works reasonably well. That means little expectancy revision is needed on our own parts. We found that policy entrepreneurs were useful but not as critical as expected. We also found that third-party intervention could be useful, but, as expected, it was not necessary either. Yet it should be kept in mind that we initially restricted our analyses to Eurasian rivalries. The reason for this is that we view Eurasian rivalries as roughly more similar in nature than the full universe of rivalries. Sub-Saharan African rivalries tend to be focused on ethnic affinities and governments interested in how domestic groups are treated by adversaries. Latin American rivalries are not totally unlike Eurasian rivalries in their preoccupations with quarrels over contested territory, but they have a propensity to endure for long periods of time with fluctuating levels of intensity. Eurasian rivalries fluctuate in intensity as well, but their lower levels of hostility never seem to dip as low as many Latin American rivalries have done.

We make no claim that these differences hinder or block the application of our termination model. Our decision to focus on Eurasian rivalries exclusively was based solely on convenience.[4] There was a possibility that the differences might matter. Equally or perhaps, truth be told, more important, we were more comfortable as authors working with Eurasian rivalries. Detailed information on sub-Saharan African foreign policy is quite difficult to attain. Spanish-language sources would be very helpful in the Latin American cases and it is a language skill none of us possess.

Consequently, we would like very much to see the model applied to Latin American and sub-Saharan African cases. There are also a number of older European rivalries (and some newer ones) that deserve analysis as well. The model was designed to be as general as feasible. Our testing has not yet caught up to the generality of the model. Hopefully, that will change. But there is also a significant exception to our Eurasian sample. Of course, we did not examine every possible Eurasian rivalry. We fully expect that there is more useful work that can be done in looking at rivalries that we chose to ignore. There is also the 800-pound gorilla of rivalry termination cases, the U.S.-Soviet Cold War, which is more Eurasian than anything else. We balked at including that rather significant rivalry in this study because it would no doubt require detailed analysis entirely on its own. The many controversies associated with disputes about what brought about the de-escalation of the Cold War are strong and the details are too numerous to deal with in a chapter or two. At the same time, we are confident that the rivalry termination model that we are advancing should work well in the case of the Cold War. It is fair to say that it would have been impossible to formulate the de-escalation model without at least thinking about the nature of the prominent Cold War rivalry. Our confidence that the model should work to explain the end of the Cold War, nonetheless, needs explicit testing.

Finally, there is one other arena of possible application that we have yet to explore. This book focuses almost exclusively on protracted interstate conflicts. There are also a large number of protracted intrastate conflicts. We doubt that our interstate rivalry model can be applied to intrastate rivalries without some modification and/or elaboration, but at the same time, we think that domestic applications are more than merely conceivable. The logic of rivalry de-escalation should be fairly generic. One way to test this assumption is to try applying the interstate model to intrastate situations.

APPENDIX

Early Theoretical Models That Pertain to Conflict De-escalation

In this appendix, we first focus briefly on listing a number of factors that have been mentioned as important to conflict de-escalation processes. Afterward, we quickly move on to the consideration of fourteen models that are relevant to rivalry de-escalation and termination. These models include Rosati 1994 and Welch 2005 on foreign policy restructuring; Checkel 1998 and Legro 2005 on ideational shifts in foreign policy; Diehl and Goertz 2000, Maoz and Mor 2002, and Cox 2010 on rivalry onset and termination; Rock 1989, T. Armstrong 1993, and Kupchan 2010 on rapprochement; Lebow 1995, Pervin 1997, and Orme 2004 on adversarial de-escalation and accommodation; and Womack 2006 on asymmetrical politics.[1] Since these models overlap considerably with our expectancy revision model, we wish to highlight their efforts in order to make the case for a partial synthesis.[2] We find that all of these models appear to be looking at similar explanations, but they privilege different combinations of factors more than others. At the same time, these models stress similar sets of variables that produce a surprising amount of overlap in the existing literature.

Students of conflict resolution maintain that the de-escalation of protracted conflicts ultimately depends on conducive background conditions that encourage adversaries to believe in settlement. Shifts in these background conditions create moments of opportunities or "ripeness" when peacemaking efforts are likely to have favorable outcomes. Table A.1 provides a list of some of the many factors that scholars have linked to the de-escalation process. It shows that de-escalation may arise from domestic conditions, international contexts, and the relations between the adversaries (Kriesberg 1991).

Past research also indicates, however, that there is no simple direct relationship between any of these factors and the termination of protracted conflict across many domains (Licklider 1993: 303). All of these variables have different sorts of impacts under different circumstances. Hence, we have no single pattern or theory of conflict resolution. In addition, Licklider (1993: 303–304) argues that conflict resolution more likely reflects a set of processes in which certain critical choice points emerge. Decisions at these points lead to alternative strategies of conflict termination. But, which situations (factors or combination of factors) present critical choice points and why is left unclear.

Table A.1. Conditions That Bring about De-escalation in Protracted Conflicts

Processes	*References*
Hurting stalemates	Zartman and Aurik 1991; Zartman 1993, 2001; Grieg and Diehl 2005, 2006; Weible and Sabatier 2006
Promise of reciprocity	Ward 1982; Touval and Zartman 1985; Goldstein and Freeman 1990; Zartman 1993; Lebow 1995; Rajmaira 1997; Goldstein et al. 1998, 2001; Bogumil 2002; Gintis et al. 2005
Extending peace initiatives	Kriesberg 1992; Hutchison 2001; Zartman 2001; Garb and Allen Nan 2006
Leadership changes and shifts in the internal distribution of power	Evangelista 1991; Licklider 1993; Bennett 1997; Mendelson 1998; Diehl and Goertz 2000; Colaresi 2004
Leadership autonomy	Stein 1993a; Coleman 1997
Third-party mediation	Touval 1975; Pruitt 1981; Becovitch, 1995; Hartzell 1999; Regan and Stam 2000; Schrodt and Gerner 2004
Mutual trust and credible commitments	Kelman 1995; Walter 1999; Bar-Tal 2000; Kydd 2005
Organizational failure and learning	Hermann 1990; Levy 1994; Stein 1995
Shifts in the external distribution of power (e.g., military defeats, loss of patronage)	Bueno de Mesquita, Siverson, and Woller 1992; Mitchell 1995; Huth 1996; Kriesberg 1998; Chiozza and Goemans 2004
Internal economic crises	Stein 1993b; Waterman 1993; Bennett and Nordstrom 2000
New perceived threats to security	Rock 1989; Lebow 1995; Thompson 2001b; Kupchan 2010

In an effort to pare down every argument that is potentially relevant to our understanding of rivalry termination, we focus on fourteen previous models of conflict de-escalation that overlap with features in our expectancy revision model. We begin with Rosati's (1994) more general model for foreign policy restructuring.

Rosati and Foreign Policy Restructuring

Rosati assumes that societies at any given point in time contain competing domestic forces that work toward and against reinforcing the status quo (Boulding 1970). Which groups gain the upper hand and in what circumstances accounts for any changes that may occur. Rosati's model concentrates on the interactions among state institutions, society, and their environment, which have an impact on the development of strategies and policies. State bureaucracies are viewed as inert institutions that move only when pushed hard to do so (Allison 1971; Halperin 1974). Meanwhile, decision makers, who preside over state policy formulation, are disinclined to alter the belief systems that underpin their choice of strategies (Holsti 1967; Jervis 1976). Mass publics also tend to be conservative because they are more comfortable with the status quo than with the uncertainties of change. Thus, vested interests and tradition pose obstacles to potential changes and leaders who advocate for such change. Concomitantly, the international environment also presents a number of built-in constraints on domestic policy shifts, including alliances, commercial relationships, legal commitments, and norms. Continuity rather than discontinuity more often characterizes the development of foreign policies and strategies (Rosenau 1981; Holsti 1982; Goldmann 1988; Hermann 1990; Volgy and Schwarz 1994).

Nonetheless, societal and environmental alterations occur and the question then becomes whether states and leaders modify their foreign policies to adapt to these new circumstances. If they choose to avoid altering existing policies, then leaders and their governments risk the increasing probability that their policies will fail, which in turn will expand the criticism from groups that advocate for foreign policy change. Domestic or international crises also open up the opportunity for such change as defenders of the status quo are overwhelmed by pressures both at home and abroad for a reorientation. In sum, foreign policy failures expose the liabilities and inappropriateness of existing strategies, shake conventional belief systems, upend government personnel supporting the status quo, and bring forth new personnel committed to different foreign policy approaches.

Rosati quotes Nisbet (1969: 282) to summarize the nature of this crisis-induced dynamic: "A given way of behaving tends to persist as long as circumstances permit. Then . . . the way of behaving ceases to be possible as the result of some intrusion, some difficulty which is the consequence of event or impact, and a period of crisis ensues."

Hence, crises create the necessary space for policy innovators and their innovations. However, there is no guarantee that such crises will lead to change. If catalysts produce positive outcomes, then leaders and governments are likely to reinforce policy continuities. If, however, crises are perceived to have negative consequences, then they are more likely to generate the quest for new policies—and the more severe the impact of negative events, the greater is the potential for foreign policy reorientation.

Based on this perspective, foreign policy restructuring tends to be fairly abrupt and episodic (Krasner 1984). Without crises, incremental change is the likely consequence. Only when the intersection of ongoing societal and environmental shifts and traditional policies combine with critical events that lead to the defeat of status quo defenders do we expect to observe a significant redirection of foreign policy. If, however, status quo defenders maintain their political control, then they are likely to support only minor or moderate alterations in foreign policy.

In sum, Rosati's model reemphasizes that true policy changes represent punctuations in an otherwise steady and stabilized course.[3] And, crises are the catalysts that overcome the normal resistance of policymakers to undertake new policy directions.

Welch and Painful Choices

While Rosati stresses the role of crises, Welch (2006) emphasizes the pain associated with the high costs of making foreign policy shifts. Welch starts with the presumption that foreign policy situations are structured to avoid major changes. Inertia is more probable than innovation. The main reasons for this bias are rooted in uncertainty, interpretation problems, and the costs of change. Welch claims that foreign policy analysts are not very good at anticipating change and that when change does emerge they are unlikely to recognize it. In addition, foreign policy organizations are slow to adapt to new environments since they are reluctant to abandon well-established routines.

Stressing the role of cognitive processes, Welch argues that policymakers are reluctant to blaze new trails. They are more likely to distort incoming stimuli than to alter their fundamental beliefs about external adversaries and threats. Leaders are also unlikely to perceive when opportunities emerge for change. And, in those cases where some policymakers observe the payoffs in trying something novel, they will assess the potential costs of pursuing them as too high. Policymakers realize that admitting their earlier policies were flawed while pursuing new policies is very risky politically. They calculate that they will be blamed personally if their new policies are unsuccessful.

Consequently, decision makers are more likely to pass up possible gains associated with potentially chancy new policies in order to avoid losses even if continuing old strategies is seen to be equally or more risky. Only when decision makers perceive that the losses associated with new policies exceed the risks of doing something different are they likely to support change. Thus, there are individual and organizational biases toward "staying the course," which ensure that foreign policy shifts will be rare and unlikely unless a disaster is looming. In this respect, leaders must be forced to change. In other words, "Foreign policy change is most likely when decision-makers perceive that their current policies are incurring painful costs; that a failure to change policy is virtually certain to result in further painful costs; and that at least one option open to them holds forth the possibility of an acceptable outcome, even if that acceptable outcome is not highly likely" (Welch 2006: 46).

Welch's argument about the inertial tendencies of foreign policy actors is compelling. Welch cites a variety of factors that work against foreign policy change: the cognitive propensities of individuals to preserve their beliefs in spite of contrary evidence; the tendencies of individuals to prefer avoiding losses over achieving potential gains; and the interests of bureaucracies to avoid innovation. Hence, the impediments to change are high and only when decision makers have their backs against a wall are they willing to pursue a new strategic tack.

Despite its appeal, Welch's argument raises questions. When do decision makers perceive they have little choice but to experiment? Certainly, there have been leaders who have preferred to go down with their ships of state—the extreme example of Adolf Hitler in a bunker in Berlin in 1945 comes to mind—than to consider more realistic (and perhaps less painful) policy alternatives. Just what does it take to compel decision makers to contemplate altering their tactics or strategies? In fact, how often do decision makers stay the course before they are replaced by different decision makers? Welch's approach seems to be tied to the likelihood that incumbents will change their minds. But, what about the possibility that turnovers of the elite are more likely to be associated with foreign policy change? How frequently do incumbents actually execute foreign policy shifts as opposed to new policymakers who are less tied to older strategies? Ideally, we would like a theory that takes into account the broader circumstances that bring about the probability of foreign policy change—as opposed to Welch's more restrictive approach on how policymakers change their own courses of action.

In a rivalry context, what prevents decision makers from escalating hostility as opposed to de-escalating it? Either course of action could constitute a significant foreign policy shift. To what extent can we generalize about whether foreign policy shifts will be more cooperative or more belligerent? Can our theoretical understanding of foreign policy change be oriented to explaining more than just a generic shift?

Checkel and New Ideas in Foreign Policy

Crises and events that can either reinforce the status quo or lead to major deviations are useful clues. But we need greater specificity in our search for ways to account for rivalry de-escalation and termination. One model closely approximates our own approach, albeit with differences. Checkel (1997) models the behavioral changes in Soviet foreign policy at the end of the Cold War on the following assertions:

1. The greater the external uncertainty or foreign policy crisis, the greater are decision makers' searches for new ideas. Thus, external uncertainty or a foreign policy crisis creates policy windows that encourage advocates of new ideas to promote them.
2. Policy entrepreneurs are critical to the empowerment of new ideas, especially in more centralized states that lack alternatives for affecting change.

3. Because centralized states lack alternatives for affecting change in comparison to decentralized states, new ideas are less likely to reach elites in centralized states than they are in decentralized states. But once decision makers in centralized states embrace new ideas, they are more likely to put them into practice and change state behavior than is the case in decentralized states.

Hence, external change and uncertainty create space for policy entrepreneurs to promote new ideas. Those states that are more authoritarian are more resistant to these cues because their systems are more closed, but they are also more readily manipulated once the new ideas are adopted. In contrast, states that are less authoritarian are thought to be more open to new ideas but less likely to implement them. Since Checkel's study focuses exclusively on an authoritarian setting, he has yet to test his ideas across centralized and decentralized states. This issue distinguishes Checkel's approach from our own.

Although we find the institutional argument to be interesting, we are unwilling to assert that institutional differences play a significant role in cases of rivalry de-escalation. That is not to say that all governments operate similarly in holding on to rivalry perceptions or that states, for instance, with electoral contests may be less susceptible to rivalry outbidding as leaders pursue votes. Rather, we prefer to remain agnostic about the role of institutions on the origins and implementation of foreign policy change at this time.

Another difference between our approaches is Checkel's focus on generic ideational change. For instance, our expectancy revision model emphasizes the shift in ideas about strategic enemies. Checkel, in contrast, is interested in ideational shifts more broadly. For instance, ideas associated with strategies of economic development may fit Checkel's emphasis on regime differences more readily than rivalry strategies. So, while our approaches appear to overlap considerably, Checkel's is couched at a more general level.

Checkel's attention to external uncertainty is a third difference. Our expectancy revision model stresses specific types of internal and external shocks while Checkel's model applies more generically to any foreign policy crisis that generates uncertainty for decision makers. We are unwilling to say that all crises are likely to have the same impact. Nor do we believe that general foreign policy crises will be the most likely settings for rivalry de-escalations, especially if the crises involve the two rivals as antagonists. In these cases, the level of intense threat is apt to discourage rather than encourage the reassessment of a rival's intentions.

Finally, we agree with Checkel's argument that changes in external environments can produce opportunities for policy entrepreneurs to lobby successfully for different approaches to solving problems. However, we maintain that the process by which policy entrepreneurs are able to gain access to or control of their governmental bureaucracies that are wedded to old strategic ideas takes time and usually occurs when foreign or domestic policy problems have become acute. Policy entrepreneurs are frequently bystanders waiting for openings to press their ideas more aggressively. Alterations in the external environment can bring about new challenges or problems for decision makers. If these alterations (either in the external or internal setting) discredit incumbents sufficiently, then policy entrepreneurs may be positioned to replace them and in that event they might be able to pursue strategic innovations.

Legro and Changing Orthodoxies in Foreign Policy

Along lines similar to Checkel's approach, Legro's approach (2005) focuses on changes in dominant foreign policy orthodoxies.[4] Although actors have interests and preferences, Legro is interested mostly in how collective beliefs or ideas guide the selection of strategies that deliver the behavior that actors want to play in world politics: integrationist, separatist, or revisionist roles.

At the heart of Legro's model, which is also critical to our own perspective on rivalry termination, is his view on external shocks. Similar to other scholars' work on foreign policy shifts, Legro maintains that shocks create windows of opportunity for change in both institutions and dominant orthodoxies. However, Legro quite rightly notes that shocks can be indeterminate. He argues that we do not know which shocks are most important in bringing about change and whether these shocks will have the same impacts across different societies. Moreover, scholars tend to look at only those cases where shocks have produced change and ignore cases where shocks have failed to produce change. Without this comparison, Legro maintains that it is difficult to discern how or whether shocks really mattered. Thus, the standard approach of "old thinking > shock > new thinking" is unsatisfactory.

Legro modifies the standard model with a two-stage argument. An old dominant orthodoxy must first collapse. In most situations, the collapse of collective beliefs is unlikely to occur. That is, dominant orthodoxies are likely to remain dominant orthodoxies for some time. How then are dominant orthodoxies sustained or swept away? Legro advances an auxiliary theory of ideational collapse that is based on four possible outcomes. If actors choose a strategy that corresponds to the prevailing foreign policy ideational framework and the results of the strategy are evaluated positively, then dominant ideas will remain. In contrast, if actors choose to pursue strategies that fail to correspond to ideational guidelines, then the results of such strategies will matter little. For instance, if the results are perceived to be negative, then the outcome will reinforce traditional ideas; and if the results are conceived to be positive, then actors will still have few incentives to challenge dominant orthodoxy. In short, "happy" outcomes are unlikely to lead to a revision in traditional ideas. However, the one situation that is most likely to create turmoil involves cases where strategies correspond to orthodox views and yet produce failed outcomes. In this context, the greater the failure, the more likely it is that actors will overthrow traditional orthodoxies.

Collective ideas do not exist without controversy since they are fought over by partisans of different perspectives. Major failures bring critics out in the open as well as shake the convictions of those who once subscribed to the dominant orthodoxy. At this point, the potential for moving to a new set of collective beliefs is greatest. But this potential must be realized by a shock that overwhelms the defenses built around entrenched ideas. The presence of a shock is not enough; the severity and the circumstances in which they occur are important as well.

In the second stage of Legro's model, new ideas must be consolidated before they become the dominant way of thinking. Consolidation, in turn, depends on primarily two factors. One is the number of alternative interpretations that exist at the time of collapse. Too many alternatives increase the likelihood that factional fighting will prevent the adoption of a new orthodoxy. In the absence of any obvious alternatives, actors will fall back on their old ways of thinking. Thus, the most favorable setting for consolidation is a situation where only one strong alternative approach emerges. The absolute failure of the old orthodoxy thereby opens up the possibility for actors to switch to an opposing viewpoint.

Consolidation also depends on how well the new way of thinking seems to be working. If new ideas appear to be failing as badly as the old ones, then they are less likely to remain dominant for long. If new ideas appear to be related to more successful outcomes, then they are likely to be retained.

Thus, there are several components in the Legro model. Shocks are prominent, but not just any shocks will do. The most crucial ones are linked to major policy failures that were expected to succeed. Then, too, these shocks only "open the door" to the possibility of a shift in prevailing orthodoxies. Partisans of alternative ideas must struggle and compete to have their perspectives adopted. New orthodoxies that emerge must be reinforced by success before they become consolidated or institutionalized in foreign policy bureaucracies. Legro's model is very similar to the one that we developed. Yet, we maintain that shocks do not always have to be linked to major failures before strategic revisions must occur. Rivals, as we saw in the chapter on India and Pakistan, are quite capable of resisting the implications of major shocks.

Rivalry Approaches to Onset and Termination

Diehl and Goertz on the Necessity of Shocks

Whereas shocks are prominent in the Legro model, they are the primary focus in the framework of Diehl and Goertz (2000). Their basic argument is that environments and interactions tend to move toward stability. In order for rivalries to begin or end, something must disrupt the stability. Hence, internal or external shocks introduce this necessary but insufficient condition. External shocks create new opportunities and expectations for rivals to initiate or terminate their conflicts, especially if the power distribution between them has been affected in a substantial way. Meanwhile, internal shocks change leaderships, regimes, and the amount of resources available for engaging in foreign policy activities.

Examples of suitable external shocks include world wars, quick changes in relative capability standings, and changes in territorial control. Primary types of internal shocks are independence, civil wars, and regime changes, especially shifts to democracy. Empirically, Diehl and Goertz find that over 90 percent of rivalry births and deaths have been associated with some kind of shock. However, their evidence also indicates that shocks are more critical for the commencement of rivalries and less so for the termination of rivalries. Hence, they conclude that shocks are a weak necessary condition for ending rivalries.

The strong emphasis on shocks is understandable. They appear to be quite common and readily associated with foreign policy changes. However, they do require careful interpretation. Should independence be viewed as an equivalent shock in kind to world wars or democratization? Often, independence does not come as a shock but is fully anticipated. It may or may not alter expectations all that much in the sense that pre-independence feuds may be continued in the post-independence era. Still, independence can significantly alter the environment in which states operate. One can no longer rely on the colonial power to police older feuds or disputes over boundaries that had been relatively suppressed in the colonial era. Similarly, democratization may reflect gradual or abrupt processes of change. It can alter fundamentally the decision-making personnel and foreign policy goals at play. So, if shocks play critical roles in rivalry behavior, then we need to ensure that their impacts on intervening events are also abrupt and unexpected and tied closely to rivalry termination processes.

Maoz and Mor

Another approach with considerable utility is advanced by Maoz and Mor (2002), who believe that they can explain rivalry onset, escalation, and termination with a relatively straightforward theoretical premise.[5] Peaceful dyads encompass mutually satisfied states, or, if there is dissatisfaction in the dyad, the dissatisfied state is too weak to do anything about it. Conflictual dyads, therefore, involve states that are not satisfied with the dyadic status quo and believe that they can do something about it.

In other words, rivals perceive that they share fundamentally incompatible goals—a perception that may be predicated on relatively accurate or inaccurate premises. Consequently, rivals are likely to engage in cycles of repeated conflict unless something changes along the lines already identified. Eventually, rivalry termination occurs when dissatisfied states become satisfied via a perceptual shift about the level of incompatibility with a rival or through the recognition that their initial perceptions were wrong. Alternatively, dissatisfied states can also transform into satisfied states when they become convinced that they lack the capability to continue contesting the status quo.

Of course, these types of circumstances may be only temporary. Maoz and Mor draw particular attention to warring rivals in which each war leaves one side triumphant and the other side in defeat. The outcome can lead to some stabilization only as long as domestic changes do not bring into power a decision-making group that is either more dissatisfied or more impressed by its capabilities than the group that lost the last war. When one of these conditions reemerges, another round of overt hostilities can be anticipated.

We also note that the outcomes of these warring rivalries since 1945 have not been all that decisive in altering the perceptions of the participants.[6] That is, winners in war are less likely to eliminate their international adversaries. One side is often defeated but its territory is neither occupied nor absorbed into the winner's expanded boundaries. Recurring conflicts between rivals may have eliminated one of the parties in earlier historical periods, but this result may be less likely in a contemporary period in which there are international restraints on the winners of war. Moreover, during the Cold War, rivals found their military capabilities rebuilt fairly quickly as a result of superpower assistance.

Although the argument of Maoz and Mor is both parsimonious and theoretically compelling, we have some problems with their approach. First, Maoz and Mor equate rivalry with conflict density. In short, rivals are those states that clash frequently and after these clashes cease, they are no longer considered to be rivals. We contend that this line of thinking indicates that Moaz and Mor are really explaining outbreaks of repeated conflict but not necessarily rivalry. Our argument is that while rivals are much more likely to become involved in repeated, overt hostilities than nonrivals, rivalries themselves can persist in the absence of explicit changes in conflict. Once we assume that rivalry can exist without overt conflict, Maoz and Mor's emphasis on satisfaction and capabilities becomes less compelling as an explanation of rivalry termination. While satisfaction and capabilities continue to be important for understanding the onset and escalation of rivalry, they are less applicable if termination is not associated with the end of overt hostilities as we contend. Hence, we believe that other variables are required to understand rivalry terminations.

Another problem with the approach of Maoz and Mor is that it is very dyadic-centric.[7] Whether rivals are satisfied or dissatisfied is an issue that is determined by the status quo that exists between members of the dyad. This may be the case for some rivalries, because dyadic members can be so totally focused on each other that little else seems to matter. However, this may not always be the case. Rivalries can be influenced by other processes that exist outside of the dyadic context. For instance, other external rivalries can influence the internal behavior of rival dyads. We maintain that an overly dyadic focus is likely to overlook the role that regional and global influences play in rivalry termination as our later case studies show.

Yet even within the scope of Maoz and Mor's theory, it is quite conceivable that satisfaction and capabilities may be contingent on events and processes occurring outside of a dyadic rivalry. For instance, the emergence of a new, mutual threat from a third party may lead dyadic rivals to reevaluate their traditional perceptions of threat. Another possibility is a shift in exogenous processes that diminish perceived threats—as in the case of the Chinese, who reassessed their threat perceptions about the United States as U.S. involvement in Vietnam declined in the late 1960s. Capabilities, too, can be directly affected by the loss of external patronage as illustrated by the Vietnamese-Chinese rivalry. Although Vietnamese decision makers did not have to deescalate their conflict with China in the aftermath of the Soviet Union's disintegration, they did have to reassess their relative capability to continue the rivalry at a fairly intense level.

Although Maoz and Mor (2002: 32–42) recognize that perceptual changes between rivals can be brought about by exogenous events, they tend to consider such factors outside the scope of their theory. Hence, a crucial difference between their approach and ours is that we believe factors often exogenous to a rivalry dyad are important to termination processes. Therefore, if external shocks are critical as other theorists argue, we want them inside a model of termination as opposed to remaining outside of it.

Cox's Rivalry Termination Model

Cox's (2010) model, unlike many other rivalry approaches, is specifically designed to account for rivalry terminations. Rivalries are perpetuated by three foreign policy orientations, which are present on both sides of the relationship. Hawks want a rivalry to continue at all costs; doves prefer to end the rivalry; and in between are the moderates who maintain the rivalry until they are no longer convinced that it is feasible to do so. Thus, the trick to bringing about rivalry termination is to reduce the influence of the hawks, increase the influence of doves, and convince moderates that the rivalry is more costly than beneficial.

These three "objectives" are more likely to be accomplished in the context of policy failures. If failures are acute and occur in both the domestic and foreign environments, then they will increase the probability of a major change in the ruling coalition. This is especially so if domestic policy failures are associated with economic setbacks. Otherwise, a leadership change is less likely if the failures are limited to foreign policy alone. Democratic institutions help to facilitate the turnover of leadership in these contexts, but they are not required. Nor is it guaranteed that hawks will fall out of favor while doves replace them; but nevertheless, policy failures create the opportunity for this to occur. In that instance, the opening for rivalry termination also emerges.

At this point, the focus shifts to the other member of the rivalry. If this adversary has also experienced domestic and foreign policy failures, then the opportunities for a change in leadership also occur. When changes in leadership occur on both sides of the rivalry more or less at the same time, the prospects of rivalry termination improve considerably. However, in the event that this does not happen, the outcome for rivalry termination will then hinge on whether the side that is attempting to change its approach offers something that the adversary is seeking. If not, then rivalry termination is unlikely. Another obstacle emerges when an adversary delays in responding to the other side's initiatives.

Cox's model is certainly testable. It is also attractive because it specifies a temporal sequence of events when bargaining could lead to rivalry termination. For instance, leadership change occurs first and is followed by a policy change; depending on the nature of these changes, they could lead to the end of a rivalry. However, the hawk-dove dichotomy in the model is awkward since decision makers may not always be easily identified as fitting into one of these categories. Presumably, the most important condition is whether decision makers who are committed to an ongoing set of policies are removed in favor of decision makers who are committed, or open, to new alternative policies. So, whether one group is more or less ideologically conservative than another is beside the point. There is also the problem of identifying policy failures, which ultimately depends on whether actors perceive them as such. One way to avoid this problem is to focus on how shocks create openings for new leadership without the requirement that they must lead to the replacement of one set of leaders for another.

Rapprochement

Rock on Peace Breaking Out

Perhaps, the earliest explicit argument for rivalry termination is Rock's (1989) interpretation of "why peace breaks out" between hostile great powers. In the first stage of his model, Rock maintains that decision makers decide that they need to abandon the war option before resolution can be achieved. Decision makers are likely to be motivated to do so when they face three conditions: 1) the emergence of a new threat; 2) the necessity of preserving economic connections; and 3) and the imperative of avoiding civil war. Once decision makers on both sides decide to reconcile, they form suitable policies in the second stage of the model. Rock argues that reconciliation is facilitated in this stage by three other factors. First, the adversaries have minimal overlap in, or commitment to, geopolitical objectives. Second, the adversaries have little economic rivalry that stems from producing and/or exporting similar commodities or lending capital in the same areas. Third, they have similar sociopolitical attributes that are likely to reduce their ideological incompatibilities.

In addition to these motivating and facilitating factors, Rock argues that a catalyst of some sort is usually needed to overwhelm the inertia associated with one side's negative perceptions of the other. Borrowing from Lebow (1981), he views acute crises as the most probable catalyst. If one side is defeated in the crisis, then that defeat may encourage decision makers to think anew about their rival. Another scenario is that defeat may weaken the influence of hard-liners or it may resolve the disputes at stake between the rivals. In short, crises and/or their outcomes encourage one or both sides to look for alternatives, and settlements that occur in the context of these crises can even generate some trust and good will on both sides of the bargaining table.

Rock downplays the role of human agents while putting greater stress on the presence or absence of key external factors that facilitate reconciliation. If these facilitating factors are present, then rapprochement is likely to occur regardless of the competence or incompetence of diplomats.[8] Thus, Rock's model has three critical components: decision-maker motivations, catalysts, and facilitating factors. Reconciliation begins when decision makers' motivations and catalysts break through the policy inertia; it is achieved mostly likely if the obstacles (or facilitating factors) line up favorably.

We agree that policy inertia must be overcome by either catalysts or shocks. An interstate crisis can serve this role, but we are not sure that defeats in crises are guaranteed to produce a reevaluation of prevailing strategies. Such defeats may lead to greater future hostility and even war if decision makers resolve to pursue a more hard-line stance in the next confrontation.[9] We believe that other types of shocks and motivations need to be considered for rivalries to end. As for Rock's facilitating factors, we are not sure whether they should be viewed as sources of rivalry or obstacles to rivalry termination. For instance, ideological incompatibilities, geopolitical conflicts, and economic rivalry are reasons for increased perceptions of threat and hostility between states. If these factors are absent initially, then the emergence of rivalries is likely to be low. If they are present, then there is no reason to presume that rivalry termination is likely to appear as an option that decision makers would consider.

We believe that it is not the presence or absence of these factors that are crucial, but whether and how the nature of these factors change over time. We argue against viewing these factors as static obstacles that must be removed for rivalries to end. Moreover, we do not agree with the notion that rivalries will persist as long as the reasons for rivalry persist. The annals of rivalry termination suggest that ideological incompatibilities, economic rivalry, and geopolitical contestation can be overcome given the appropriate circumstances—as illustrated by one of Rock's motivating factors (the emergence of a new and more serious threat) for instance.[10] Consequently, we are not convinced that "facilitating" factors necessarily trump "motivational" or catalytic factors. We also maintain that reciprocity and reinforcement are critical facilitating factors in the process of rivalry transformation. Despite the shortcomings in Rock's model, we draw on a number of its arguments to inform our own expectancy revision model.

Armstrong

Tony Armstrong (1993) presents another set of pertinent clues for rivalry termination.[11] Although he (1993:134) concentrates on negotiation tactics that two rivals employ once they explore the prospects of rapprochement, he offers two generalizations about the "international circumstances" that will encourage both sides to come to the table in the first place. He maintains that improved relations are more likely to succeed when the following conditions are present for both governments: 1) the general strategic and/or diplomatic situation threatens to deteriorate and continued hostility obstructs the governments' ability to address the problem; and 2) decision-maker expectations about the nonviability of their adversary's regime are replaced by expectations of the viability of the regime.

The first condition relates to environmental deterioration that implies an implicit invitation to contemplate a change in strategies. Armstrong focuses on perceptions of impending deterioration as opposed to Rosati's realized crises. Yet both of Armstrong's conditions introduce the idea of expectations—albeit not very explicitly. His first condition speaks to expectations of strategic deterioration. His second addresses expectations about an adversary's viability in a context of abandoning earlier hopes that the other side might solve one side's problems by conveniently collapsing. While this factor does appear in some cases, it is not clear that it shows up in enough cases to make it a compelling explanatory element. Still, one could argue that Armstrong's conditions are anchored clearly in decision-maker expectations that encompass perceptions concerning more generally the state of the environment in which they operate and more specifically the status of the enemy rival.

Kupchan's Stable Peace Model

Kupchan (2010) embeds rivalry termination within a more elaborate stage theory designed to explain stable peace outcomes. Since stable peace goes far beyond rivalry termination, our current interest is focused primarily on the initial stages. The first phase is centered on one side's unilateral accommodation. When a state is beset with multiple threats, it is logical for that state to attempt to reduce the number of threats with which it is coping by selectively signaling its benign intentions. If concessions can work to eliminate one adversary in a crowded field, then the accommodating side is likely to have improved its general security position, at least if it is successful.

The first phase dampens the selected rivalry. A second phase furthers it to the extent that both sides become more accommodative. Gradually, the rivalry disappears and opens up the possibilities of the third and fourth phases, which allow for increased interaction and the formation of communal identities between the two states. Whether all four phases will come about depends in turn on institutional restraints and compatibilities in terms of social orders and culture. Domestic restraints on the exercise of power facilitate the communication and imagery associated with relatively benign intentions. More essential, however, are compatible social structures in which the distributions of power among various types of domestic groups are not threatened by more peaceful interactions with the former adversary. To generate new narratives of common identity, it is also critical to be able to draw on a common cultural heritage.

Thus, Kupchan begins with limited rapprochement (phase 1), which can broaden into something like a security community if the appropriate ingredients are in place. For our purposes, however, this model resembles most closely the British-U.S. transition from rivalry to a special relationship in the twentieth century. In that respect, it narrows its broader rivalry applicability by specifying a single path—unilateral threat reduction—as the termination trigger. The desire to reduce the number of existing threats does arise from time to time, but it is not sufficiently widespread to explain a large class of rivalry terminations. Something more general is needed to encompass multiple triggers for rivalry termination.

Adversarial De-escalation and Accommodation

Lebow on Adversarial Accommodation

A different approach, although one that certainly agrees with Orme's emphasis on leadership skills and choices (discussed below), is taken by Lebow (1995:181), who hypothesizes that conciliatory accommodation between two adversarial states is most likely when leaders 1) are committed to internal reforms that require or are expected to benefit from improved relations with an adversary; 2) recognize that confrontation has been counterproductive in the past; and 3 expect their adversaries to respond positively to conciliatory overtures.

Lebow refers to these conditions as propositions about alternative pathways to accommodation. He believes (1995: 182) that other pathways also exist and suggests that mutual fear of a third party and economic incentives may also play some role. In a subsequent article (Lebow 1997: 165) he adds a new condition relating to the domestic feasibility of mobilizing support for conciliatory foreign policy. Moreover, implicit to both of his treatments is a very strong focus on the leaders involved. The ending of the Cold War was very much a Gorbachev story. The de-escalation of Egyptian-Israeli tensions was very much a Sadat story, and so on. Yet in both analyses, Lebow (1997: 175) also makes sure to stress that his hypotheses "represent a tentative step toward a more comprehensive theoretical explanation . . . and need to be incorporated into a broader theory that more fully specifies the . . . other conditions."

The expectancy theory that we have put forward is intended to approximate that broader theory to which Lebow refers, even though it may not be exactly what he had in mind. A variant on his first condition (commitment to domestic reforms) is incorporated as one of several sources of shock (Rosati's crises)—as are his suggestions of third-party threat and economic incentives. His second condition does not appear in the theory in any explicit way (essentially, it is taken for granted that innovative policy entrepreneurs are dissatisfied with the old ways of doing things), but his third condition (expected reciprocity) is one of the main components. The idiosyncratic element of leadership is translated into a strong version of policy entrepreneurship. Finally, his fourth condition (domestic support) has a role to play as one source of continued reinforcement for what is seen as a gradual process of de-escalation once such a course is set in motion.

Pervin on the Transformation of Antagonism

David Pervin's (1997) perspective on the transformations of antagonisms in the Middle East presents a different type of challenge. He begins with a set of assumptions that merge into something that approximates a compact and widely applicable theory. Decision makers are rational actors who pursue the primary goal of staying in power. But to achieve this goal, decision makers must have access to the resources that make success possible. Whatever affects access to the necessary resources will force decision makers to reconsider their strategies. Whether new strategies are likely to be forthcoming, in turn, depends on a variety of other factors. Policy alterations require that inertia be overcome, and the greater the previous investment in a conflict situation, the greater may be the sunk costs that must be overcome or ignored.

Whether and how policies can be altered also depends on how difficult it is to obtain resources from domestic and international environments. The greater the ease of access to resources, other things being equal, the less likely is policy alteration.

Another factor is the symbolic value of the conflict. Minor conflicts are more easy to overcome than are conflicts that can have a large number of significant repercussions. For instance, there may be an explicit dispute over who is allowed to control adjacent territory. Territorial issues can be sufficiently intractable in their own right, but they become even more awkward when surrender might have implications for a state's regional leadership position. Implications such as these compound the costs of concessions. Then, too, concessions can have variable impacts on domestic actors. Internal political and economic changes may be deemed too expensive to salient constituencies. Finally, still another consideration concerns how long prevailing circumstances are likely to persist and what might follow in their wake. If the current international environment is thought to be relatively long-lived, or if the future alternatives are even less desirable, then decision makers may be more likely to adjust their strategies.

What one should look for, then, are actors adapting their strategies to changing environments, subject to a number of caveats. It is not just any change that matters but, rather, ones pertaining to the survival of decision makers and the resources essential to that goal. Pervin illustrates his perspective by a brief examination of foreign policy changes in the relations between Jordan, Syria, the PLO, and Israel in the mid-1990s. The environmental changes were multiple and sequential. The Cold War between the United States and the Soviet Union had ended, leaving the United States the primary surviving superpower in the Middle East. Soviet allies could no longer count on being rearmed after losing a war with Israel. Arab states could no longer play the two superpowers against each other. The general effect was to make interstate war less likely and Arab actors less independent of external patrons. Another change was the second intifada, which made the perceived costs of Israel holding on to territory in the West Bank indefinitely more costly.

A third major change was the Iraqi invasion of Kuwait and the eventual defeat of Iraq by a large coalition led by the United States. The salience of the United States was reinforced while Jordan and the PLO found themselves more or less on the losing side of the Persian Gulf War. Palestinians were expelled from the Gulf area, putting pressure on Jordan to absorb them despite major domestic economic problems while the PLO's funding was cut severely. In addition, the PLO was facing an increasingly serious challenge from Hamas over the leadership of Palestinian resistance to Israel. Jordan was being pressed by the United States to improve its relationship with Israel in return for debt relief and arms.

Syria, in contrast, had lost its Soviet patronage but gained from being part of the winning coalition in the Persian Gulf War. Its domestic economic problems were much less severe than Jordan's and, unlike the PLO, the Syrian government had no domestic challenger. Its main territorial issue with Israel was difficult to approach incrementally. Either Israel would return the Golan Heights or it would not. To concede on this issue would not only be extremely unpopular at home, it would also reverberate throughout the region, negatively affecting Syria's claims to Arab leadership.

Which of the three Arab actors was most likely to reduce its conflictual relationships with Israel? The theory predicts that Jordan and the PLO, as the most vulnerable actors, would be more likely to increase cooperative interactions with Israel than would Syria, the least vulnerable of the three. This is precisely the outcome that developed in the mid-1990s. Israel and the PLO negotiated the Oslo Accords. Jordan and Israel negotiated an end to their interstate rivalry. Little change took place in the Israeli-Syrian relationship. Of course, Israel need not have been receptive to cooperative overtures, but the overthrow of Likud's rule by Labor made receptiveness far more likely. Israel's economic future was also seen as benefiting from a substantial decrease in conflict. Opportunities for investment and technological production could be expected to expand in a less risky setting.

Pervin greatly (over)simplifies the increased cooperation in the Middle East by contrasting three preference schedules.[12] One "Arab" preference schedule is compared to two Israeli (Likud and Labor) sets of preferred futures.[13] Pervin's basic point, however, is that the Arab and Likud preferences had been vitiated by environmental changes. Israel was too strong militarily and the Arab actors too weak militarily to destroy Israel. Without the fallback support of Soviet patronage, Arab-initiated warfare was both unlikely and too dangerous. The intifada had made holding on to the occupied territories gained in 1967 less attractive than had been the case earlier. As a consequence, the Arabs were reduced to their third and fourth preferences (peacefully regaining or conceding the territories). The defeat of Likud by Labor meant a preference shift within Israel, brought on by the Labor schedule appearing more attractive than the views held by the incumbent regime.

The ascendancy of Labor meant that Israel was prepared to negotiate peacefully over the status of the occupied territories at the same time that at least some of its Arab opponents were also prepared to be more cooperative than in the past.

Pervin goes on to note that the longer these environmental conditions persisted, the more likely it would be that Middle Eastern relationships would be transformed. The key is for the international system to create incentives for cooperation. The longer these incentives remain in effect, the more solid the ground for gains through cooperation—especially if important domestic actors view their interests as being best served by the maintenance of pacific behaviors and environments.

We have no problem with arguing that environments, prevailing preferences, and inter-actor relationships evolve or change over time. When environments change, actors are likely to need to reconsider whether their leading preferences and strategies need to change as well. The end of the Cold War, the intifada, and the Persian Gulf War certainly constitute or reflect environmental changes in the Middle East. We emphasize the same types of factors but call them "shocks" instead of changes for they are not simply gradual changes but rather abrupt ones that often catch the actors involved by surprise.[14] As shocks, they represent opportunities for reconsidering strategies toward adversaries, among others. To the extent that the responses to the shocks lead to behavior that is more in tune with the altered international environment, actors can be said to have adjusted or adapted their behavior. In this respect, we are not really in disagreement with Pervin's approach. Rather, we are emphasizing different aspects of the same environmental changes.

We do not feel the need to assume that decision makers are rational actors engaged in careful cost-benefit calculations while also primarily seeking to remain in power. Real world examples of behavior that conform to these assumptions are not difficult to find, but they are not critical to the building of our theory. We do accept the significance of changes in decision makers' access to resources encouraging revised calculations. Shocks are frequently shocking because they impede or cut off access to resources that decision makers had come to rely on in earlier times.

We certainly have no reservations about accepting the utility of preference schedules. Political actors rarely fixate on one possible eventuality as the sole acceptable outcome. Instead, they tend to identify more and less favored outcomes. They may work harder to attain the more favored outcomes, but sometimes they must accept the less favored ones. Alternatively, actors who are fixated on prevailing preferences that are no longer seen as viable may find themselves ultimately removed from office.[15]

One of the more interesting aspects of Pervin's argument is the implication that images of adversaries never change. Preferences may have to give way to altered realities from time to time, but the idea that decision makers' perceptions of their adversaries and adversarial intentions might change and then generate new slates of preferences and strategies is entirely absent. We think this assumption is too strong, but it does highlight an important dimension of analyzing rivalry terminations. If one emphasizes the perceptual/psychological foundations of rivalry, then it is not difficult to construct a model of rivalry termination that would be predicated on decision makers realizing that their opponents were not quite as malign as they had previously thought and, therefore, more suitable for cooperative exchanges. Pervin, no doubt, would reject such a model as naïve and implausible. In contrast, we do not rule out such a possibility, but we do not find it all that probable. We think policy about rivals is based on expectations about what strategies are most likely to work, given the adversary's capabilities and intentions and one's own capabilities and preferences. A more benign view of an adversary could lead to altered expectations. But so too could perceptions of an adversary's weakness or strength, an adversary's apparently altered strategies, and one's own relative weaknesses or strengths that would also encompass changes in opportunities and threat perceptions.

Accordingly, decision makers do not have to learn to love their adversaries in order to ratchet down interstate tensions and rivalry relationships. Threat perceptions, for instance, are malleable. They can be exaggerated for conscious effect and they can be downplayed when convenient. One illustration is the U.S. reaction to the importation of Soviet missiles in Cuba in 1962 and its much more low-keyed reaction to information on the development of Soviet submarine bases in Cuba less than a decade later. Neither of the reactions necessarily implied that U.S. decision makers had drastically altered their image of Soviet decision makers and their intentions. Similarly, the negotiations involving Nixon, Kissinger, Mao, and Zhou in the early 1970s did not hinge on abrupt U.S. or Chinese reappraisals of U.S. or Chinese benignness or intentions. The beginning of U.S.-Chinese rivalry termination was instead predicated primarily on calculations of the utility of reduced U.S.-Chinese hostility vis-à-vis U.S.-Soviet and Sino-Soviet relations.

We see such changes in interstate relations as something more than merely moving down one's collective preference schedule. But we also think that agents are often necessary to the process of changing preferences. That is, preference adjustments do not happen automatically. They require policy entrepreneurs who may or may not be able to manipulate their bureaucratic resources toward new ways of doing business with adversaries.[16] Thus, Hubert Humphrey, if he had defeated Nixon, might not have been able to, or even likely to, seize the opportunity that Nixon and Kissinger exploited in 1972. That does not mean Nixon was necessary to the U.S.-Chinese rivalry termination process, but he was certainly a part of it. Much the same statement could be applied to Rabin in the Israeli-Palestinian negotiations of the mid-1990s. His ascendancy to the Israeli prime ministership and his subsequent removal from that office were not marginal to the success or failure of the Oslo Accords. Policy entrepreneurs can be an important variable in altering preference schedules.

Finally, we agree that time is critical to rivalry transformation. Decisions to de-escalate antagonisms are not like light switches that can be turned on and off. Rivalry de-escalation processes are begun at some point and then take time to gel into more concrete relationships. But more than time is required. The initial decisions to alter an adversarial relationship need considerable reinforcement. Both sides need to feel that they have not made a fundamental error in judgment. The identity of the policymakers involved may change, but their behavior toward the former adversary must remain roughly consistent with expectations of reduced threat. Otherwise, it is only too easy to slip back into the conventional groove of hostility between rivals. This reinforcement process is also a critical part of the rivalry termination process.

Orme on Rivalry De-escalation

Orme's (2004) interpretation of rivalry de-escalation is similar to Armstrong's. Orme believes that analysts give too little credit to statesmen. It is their task to monitor an adversary's intentions and capabilities and to respond as best they can within the constraints imposed by levels of domestic support for various foreign policies. If the issues that separate rivals are intractable, then there may be little recourse but to fight for one side's interests.

Where issues are tractable, however, some possibility of negotiating a noncoercive outcome is possible. A variety of factors may encourage or discourage successfully negotiated cooperation. Leaders may feel, for instance, that their political survival hinges on the continuation of external conflict. If so, then genuine cooperation with external adversaries would seem to be unlikely. Proliferating external threats and economic stagnation, however, can create an atmosphere of pessimism that may facilitate negotiation.

It is pessimism that Orme anoints as the key to peace between rivals. As long as rivals believe that they are either stronger than their adversaries or that there is some good chance that their own capabilities will improve in the near future, negotiated accommodation is improbable. Intransigence, resolve, and/or outwaiting the enemy seem most advisable. If one rival thinks its own capabilities are weaker or unlikely to improve, then some type of accommodation may become attractive—both as a leadership goal and a strategy that can be sold to domestic audiences. In this respect, Orme believes that pessimism is more important than distrust between rivals. Given sufficient pessimism and effective leadership, the distrust can be overcome. In the absence of substantial pessimism, few incentives exist to make any concessions.

Orme may be right about pessimism. It is hard to imagine optimists making substantial compromises. One or both sides need to feel a need to try something different and break out of an international relationship that appears to be thwarting the attainment of other goals. It seems quite possible that deteriorating military and economic capabilities, as Orme suggests, can encourage some decision makers to reassess their foreign policies toward traditional foes. But, as in the case of Maoz and Mor's (2002) approach, Orme's focus is primarily on the psychological mood within one or more rivals. Moods can be rather tricky to operationalize. Whose moods count? How ephemeral are decision makers' moods? Are they up one week and down the next? One way to look at what we are doing is to say that our question involves what triggers more pessimistic outlooks and, once triggered, what else is necessary to bring about rivalry de-escalation.

Asymmetrical Politics - Womack

The final model to consider is Womack's (2006) asymmetrical politics model. Womack begins with the assumption that asymmetrical relationships between states work differently from symmetrical relationships. Symmetry allows both sides to contemplate equivalent moves. Asymmetry means that some significant moves are foreclosed to the weaker side. For instance, the stronger side might be able to invade and conquer a weaker opponent but it is most unlikely that the weaker opponent could do the same. As a consequence, asymmetrical relationships will tend to be susceptible to relative inattentiveness by the stronger side and overattentiveness by the weaker side. The stronger side will tend to focus on relationships with states that are more competitive while the weaker side may become overly sensitive to interpreting the stronger side's actions and intentions.

A second theoretical characteristic of asymmetrical relationships is differential expectations. The stronger side expects the weaker side to show deference to the stronger side's greater capabilities. The weaker side expects the stronger side to respect its own autonomy. Misperceptions can arise on one or both sides, but the chances for conflict are especially magnified if the weaker side ceases to be deferential after perceiving the stronger side to be threatening its autonomy. It is also hardly unknown for the weaker side to exaggerate its own capability to withstand attack from stronger parties. What may thus be viewed as a policy nuisance initially by the stronger side can be seen as a matter of national survival by the weaker side.

Should the stronger side attempt to bring the weaker party to heel coercively, it is unlikely to mobilize its full capabilities. The degree of explicit threat simply is not there for an all-out attack. The stronger side probably also has stronger enemies to worry about. Should the weaker side muster sufficient resistance to thwart the stronger side's punitive attempt, a stalemate may ensue in which neither side is likely to win. The weaker side has too much to lose to surrender and can utilize asymmetric warfare techniques to its own advantage. The stronger side can keep the stalemate going indefinitely but risks losing prestige, resources, and domestic support for the effort if the stalemate is allowed to persist indefinitely.

When both sides conclude that the stalemate is unlikely to work in their favor and that a less hostile relationship is preferable to the status quo, negotiations over normalization are conceivable. Then the question becomes whether some new framework of deference for the stronger side and autonomy for the weaker side can be constructed as negotiators consider the terms of resolving outstanding policy differences. The movement toward a "normal" relationship is not likely to be immediate because suspicions on both sides must be first overcome. But if the normalization process appears to be working, decision makers on both sides may become convinced that the new relationship can work. Until that point is reached, though, some probability of intermittent threats and crisis lingers on. Should the status of normalcy be attained, however, the new relationship is apt to be more stable than was the case prior to the development of active hostility because both sides are more likely to avoid some of the misunderstandings that led to the earlier escalation of conflict.

In Womack's worst-case scenario for asymmetrical relationships, something leads to conflict escalation and an attempt to resolve the issue by the stronger side through the use of military force. If the application of military force fails to achieve the desired outcome and the weaker party is able to resist successfully the stronger side's efforts, then a state of stalemate sets in. Of course, nothing guarantees that the stalemate will not persist indefinitely. At least one side, and preferably both, has to realize that the stalemate is less desirable than yielding on policy differences. At this point, some potential for negotiation and normalization of the relationship is possible. Normalcy is attained when both sides reach the point that they can assume that policy differences are likely to be negotiable.

One of the definite pluses of Womack's model is that it was developed to conceptualize processes in what we view as an asymmetrical rivalry between China and Vietnam. It offers, therefore, a rivalry model with some attractiveness even though Womack does not use the term rivalry. The twin theoretical features of over/under attentiveness and defense/autonomy deserve further application to other asymmetrical conflicts, such as that between Russia and Finland or the United States and Cuba. Womack's perspective also emphasizes the expectations of the antagonists—a perspective that we showed some appreciation for in Chapter 2. The stress on the slowness of normalization processes is also highly appreciated. Yet the basic nature of a model explicitly developed to deal with asymmetrical situations limits its applicability for rivalry purposes. Asymmetrical rivalries have existed and some have continued much longer than one would expect. Still, they constitute a definite minority in the rivalry population.

The question remains whether we need to focus solely on models for specific types of situations or whether it is possible to develop more general conceptualizations that can be applied to both symmetrical and asymmetrical rivalries. At the same time, we are not suggesting that researchers must choose one or the other approach. Limited models can have great utility for the populations that they address, especially if they are fairly consonant with more general models. We think the model developed in Chapter 3 is quite compatible with Womack's useful interpretation.

Summary Observations on the Other Approaches

Thus, there are a few points that can be made based on this very selective sampling of the literature on, or pertinent to, rivalry de-escalation. One is that earlier efforts have been content, for the most part but certainly not entirely, to advance isolated hypotheses (see, for instance, Table A.1) or very general models. The need for a broader and more comprehensive treatment has not gone unrecognized. Another is that the substantive arguments relied upon in earlier efforts need not be abandoned. No effort need be made to incorporate every possible proposition, but it is quite likely that a more comprehensive yet explicit theory will overlap substantially with earlier hypotheses and models. The difference is that the theory needs to tie the components together in some coherent fashion. It also needs to generalize as much as possible the conditions that are thought to influence efforts at de-escalation. That may mean using different language to convey (and to cover) types of behavior similar to those discussed in earlier propositions. Finally, the sampled literature by Rosati, Checkel, Legro, Maoz and Mor, Armstrong, Orme, Welch, Lebow, and Womack highlights, without exception, the standard mode of analysis on rivalry de-escalation. Things change in decision makers' environments that lead them to revise either their view of the adversary or the wisdom of their existing strategy for coping with the adversary. This theme constitutes the core of the expectancy theory developed in this book. Changing decision makers' minds is critical and necessary to creating situations in which de-escalatory negotiations between rivals can begin—a point made by Armstrong, Orme, and Womack. But, we have argued, these changes are not sufficient to bring about rivalry de-escalation.

We have proposed that an expectancy framework is best positioned to frame the types of processes upon which we are focusing. It offers a viable means of weaving together a wide range of seemingly disconnected variables into a coherent explanation of de-escalation. We maintain that protracted conflicts are processes that are subject to change in the context of environmental challenges. In other words, protracted conflicts are sustained by the strategies and policies of adversaries, their expectations, and their institutions. Over time, these strategies are reproduced in a routine fashion that produces inertia.

Protracted conflicts can be seen as partial equilibrium systems. They exhibit stability over time, but during certain periods the possibilities emerge for rapid, dramatic, and nonincremental change in the level and intensity of the conflict. A key property of protracted conflicts, therefore, is their inertia—the point at which the adversaries' expectations and behavior are said to be in equilibrium. In other words, the continuity (or stability) of protracted conflicts reflects the entrenched expectations that each side has of the other in terms of their tactics, strategies, and past actions.

Something has to give for fundamental change to emerge in the relations between two adversaries. We think protracted conflicts de-escalate when adversaries adopt new interpretations, understandings, and expectations of their opponents (see Adler 1991). When is such innovation most likely to occur? We think that crises or shocks are the critical factor in bringing about the realization of alternative expectations. Periods of stasis or equilibrium are eventually interrupted by shocks or incidents that tip the level and intensity of the protracted conflict into a new direction that can lead either to escalation or de-escalation (Young 1998: 12–15).

Shocks consist of an event or a short-term chain of events that are potentially major turning points in political behavior. They can be either exogenous or endogenous, although neither is assumed to have a more important effect than the other. Exogenous shocks emerge from the environment that is external to the protracted conflict (as in the cases of changes in international distributions of power or war). Endogenous shocks occur within the domestic contexts of the adversaries (as in economic crises).[17]

Shocks have both subjective and objective dimensions (Knight 1998). The objective dimensions are the event or series of events that occur abruptly and have the capacity to bring about a change in behavior. The subjective dimensions deal with the orientations, attitudes, or perceptions of the adversaries involved. If this interpretation is correct, then adversaries are unlikely to change their policies unless shocks induce the realization that existing strategies (based on outmoded expectations) are infeasible or potentially disastrous if they continue. Only then does cooperation and compromise become possible.

Political shocks, as a consequence, are necessary conditions for the initiation and consolidation of de-escalation. They are transforming events that cause adversaries—or, more likely, some elites within the antagonistic rivals—to adopt new ways of thinking. When adversaries realize that the strategy they had been pursuing cannot triumph or they cannot gain more by continuing it, a more accommodative strategy may offer a better alternative (Kriesberg 1998: 217). Therefore, shocks can change the expectations and strategies of at least some of the critical participants who then attempt to undertake new policies that move intractable conflicts into seemingly less intractable ones.

Yet shocks are at best necessary but not sufficient triggers for de-escalation. They are not sufficient because shocks do not always result in initial accommodations, especially if they fail to elicit new ways of thinking in one or both of the adversaries. Moreover, shocks may have long-term impacts as result of their intermediary effects on other internal or external processes that more directly bring about changed expectations. What these shock-induced changes are likely to be are hard to predict given the variety of events and circumstances surrounding protracted conflicts. Hence, it is impossible to guarantee when and where shocks will lead to political breakthroughs.

What events qualify as shocks with the potential to de-escalate hostilities? Unfortunately, the list of possible candidates is long and diverse. However, recent empirical research by political scientists and sociologists on social movements, revolutions, regime transitions, and elite settlements point to the role of certain types of recurring events: defeat in warfare, economic crises, withdrawal of foreign support, and the loss of societal support for existing policies.

While military defeats and/or the withdrawal of foreign support have typically been associated with revolutionary situations, state breakdowns, and crises within the elite (Tilly 1978; Skocpol 1979; Leng 1983; Reiter 1994; Goertz and Diehl 1995, 1997; Goldstone 1991; Oberschall 1996; Dogan and Higley 1998), they also play a critical role in de-escalating protracted conflicts. These events affect the capacity of adversaries to mobilize the necessary resources to sustain protracted conflict. Diminished resources (in terms of economic and/or political support) may force parties to reevaluate the conditions under which they are willing to seek an initial agreement (Lichbach 1995: 167–209).

A sharp deterioration in a country's aggregate economic performance (including the perception that economic conditions will continue to worsen) can threaten the political survival of ruling elites unless they can reverse the situation. Ruling elites who face challenges from their opposition groups and mobilized publics are likely to reverse policies that are construed as obstacles to economic recovery (Haggard and Kaufman 1995; Dogan and Higley 1998). Consequently, leaders will reevaluate their willingness to prolong conflicts that drain important societal resources or block needed reforms for long-term economic growth (Bunce 1981; Evangelista 1991; Licklider 1993; Lebow 1995; Bennett 1997). Stein's (1993b) analysis of the bargaining environment regarding the Camp David Accords of 1979 provides a good example. She argues that acute political and economic problems in both Israel and Egypt (although more intense in Egypt) changed the bargaining strategies of Sadat and Rabin, both of whom sought a political agreement that would reduce not just their security threats abroad but would pave the way for more foreign aid and economic liberalization at home.

One reason why protracted conflicts last as long as they do is because they are sustained by what Baumgartner and Jones (1993) call "policy monopolies" within one or both of the adversaries. Policy monopolies reflect dominant political understandings about a policy issue—akin to Legro's dominant orthodoxies—in this case the conditions surrounding interactions and negotiation with an adversary. They are reinforced by existing institutional structures (which limit access to the policy process) and powerful supporting ideologies that reside within these institutions. These buttressing ideologies are usually connected to core political values, which are communicated directly through images and rhetoric to the public. Policy monopolies last until they are supplanted by new political and/or social movements that are predicated on new ideas that become popular and diffuse throughout the population (Baumgartner and Jones 1993: 6–16)—at least in more democratic settings. In less democratic settings, it may suffice to supplant policy monopolies exclusively within elite circles.

Consequently, internal uprisings, civil strife, or protest cycles (Burton and Higley 1998; McAdam, Tarrow, and Tilly 1997) may reverse policy monopolies by either replacing old institutions and/or elites with new ones or encouraging old elites to reevaluate their modes of thinking. Defeat in war and economic crises may also lead to these sorts of changes as well. In any case, new expectations can emerge, which may lead to policy changes vis-à-vis external adversaries.

To the extent that shocks or crises hasten the re-evaluation process, some actors (certainly not necessarily all) move from one set of collective understandings to another. If the direction of the shift produces negotiations, then these collective understandings will be communicated through diplomacy. That is, leaders will express their new understandings to their opponents, which should facilitate the process of negotiation and conciliation (Adler 1991). Cooperation and de-escalation, therefore, are rooted in the strategies of political survival. How, why, and which environmental crises are likely to bring about escalation or de-escalation was discussed in more detail in Chapter 2.

Shocks can induce indirect changes as well. Disasters and crises reinforce preexisting perceptions of a problem among elites and the public alike. If people perceive that existing policies are not meeting certain goals, that the costs of these policies are escalating, or that they are generating unanticipated negative consequences, then public support for ruling elites and their policy monopolies or orthodoxies may erode. Shifts in public opinion, for instance, can lead to administrative or legislative turnovers from which policy entrepreneurs are likely to emerge. By their very definition, policy entrepreneurs—actors seeking political gain (and improved policy) by innovating new ways of treating public problems—bring about the destruction of policy monopolies by advocating new solutions to old problems and selling new proposals to elites and political and/or social movements. Most of all, they exploit political events and opportunities to differentiate their positions on pressing policy problems (Kingdon 1995: 182). They are political pioneers, willing to venture into new terrain by pursuing new solutions to old problems.

Yet, policy entrepreneurs are hardly passive actors waiting in the wings for the appropriate constellation of factors to appear. They are apt to be proactive. They seize opportunities to carve out more political space that makes policy change possible. Shocks can help generate these opportunities, especially if they have produced major structural breakdown, fragmentation, and dissension. Entrepreneurs are then in a position to restructure their political environments in ways that provide them with enough political power to overcome internal commitments to older, more entrenched policy strategies. In short, policy entrepreneurs have to have more than new ideas. They must have sufficient political resources to support and sustain new strategic overtures toward cooperating with their adversaries.

Finally, they need reciprocal gestures from their adversaries lest their efforts to promote greater cooperation resemble the act of one hand clapping. Decision makers may also need pressure and encouragement from third parties. Third parties can also sometimes smooth the way toward greater cooperation. Moreover, the cooperative sequences, once instituted, must continue to be reinforced by the actions of both sides (Burton and Higley 1998). If not, then there is always some likelihood of one or both parties reverting to their default position of protracted conflict. Thus, post-settlement shocks may either derail the process or reinforce the commitment by elites to abide by the initial accommodation. Conciliation and cooperation, therefore, can be as protracted as the initial conflict seemed.

We contend, in sum, that shocks, expectation revisions, policy entrepreneurs, reciprocity, third-party pressure, and reinforcement can be critical to the de-escalation and termination of rivalry relationships. We do not expect each and every factor to be equally important to changing strategies. The questions are whether the theory is helpful and whether the various factors specified by the theory are of equal importance.

NOTES

Chapter 1

1. Gaddis (1992/1993) once even castigated international relations theory for failing to predict the end of the Cold War. It would be a good story if that scolding led to more explicit analyses of rivalry, but there is no evidence that a cause and effect relationship exists. Still, if we had known more about rivalry termination processes in the 1980s, the end of the Cold War might have come as less of a surprise.

2. Strategic rivalries are defined as relationships between states viewed as threatening competitors and, therefore, as enemies (Thompson 2001a; Colaresi, Rasler, and Thompson 2007). Such rivalries can be identified by consulting diplomatic histories and decision-maker discussions pertaining to their country's primary adversaries. This approach to operationalizing rivalry should not be confused with an alternative approach to identifying rivalries in terms of the frequency of militarized interstate disputes (MIDs) that occur within specific time periods (see, for instance, Diehl and Goertz 2000). One basic problem in differentiating the two approaches is that only some strategic rivalries participate in a sufficient number of MIDs to qualify as rivalries from a dispute density perspective. At the same time, some dyads with a relatively large number of MIDs do not necessarily qualify as strategic rivalries. Decision makers on one side view the other side as a nuisance or even a problem that falls short of a threatening enemy. Thus, choosing between the two approaches hinges on what one hopes to capture in an analysis.

3. Rivalries were coded for the primary reasons analysts gave for termination outcomes. The multiple types of outcome explanation were then condensed into the three generic types enumerated in Tables 1.1–1.3. In some cases, no explanation was found. These cases have been placed in the negotiation category identified in Table 1.3. In other cases, multiple explanations were given for a single case. When this occurred, an effort was made to focus on the single argument that seemed most prevalent or significant.

4. Any one of the four paths, of course, may well involve negotiations. The fourth category is one in which termination or significant de-escalation occurs after some negotiation but in the absence of the first three types of circumstances. In general, the approach adopted in this study black-boxes or ignores the specific nature and content

of negotiations. This decision is not intended to imply that negotiations and negotiators are irrelevant to the outcome but only to suggest that the theory focuses on the context in which negotiations are most likely to ensue. It also seems fair to acknowledge that it is assumed that negotiations alone are insufficient for rivalry termination—that is, they are unlikely to succeed in the wrong context. That does not imply that negotiations will always succeed in the right context. The expectancy theory relied on in this analysis, then, can be said to focus on the most propitious context for successful negotiations.

5. The theory explicitly evades a commitment to any of the more well-known paradigms in international relations (realism, liberalism, constructivism, Marxism) by combining elements from several different paradigms. There is also little consideration given to mass attitudes that are assumed here to lag elite attitudes toward external rivals. At the same time, it is acknowledged that strong popular hostility toward external rivals can be manipulated by domestic politicians to block or slow accommodative strategies proposed by other domestic politicians.

6. Policy entrepreneurs and third-party pressures, we argue, can be useful but are not absolutely necessary.

Chapter 2

1. We dispense with any conventional review of the literature and move directly to outlining our argument. An extensive review of the literature, however, is offered in the appendix. Readers who want the review can stop at this point and turn to the appendix first before reading Chapter 2. Readers who do not want the literature review can choose to ignore the appendix. Nonetheless, our contention is that the theory described in this chapter overlaps in a number of ways with earlier, often more narrow approaches. Evaluating that claim requires reading the appendix at some point.

2. This theory relies heavily on the argument found in Young 1998 but also overlaps extensively with the literature noted in Rasler 2001b, as well as with earlier and alternative explanations of rivalry termination discussed in the appendix. Obviously, interest in this question of rivalry de-escalation and termination remains quite healthy.

3. A prominent woman conductor was once asked if being a woman caused her problems in leading an orchestra. Her reply was that she couldn't be sure because she had never been a male conductor. This is a sophisticated response suggesting some sensitivity to analysis of variance problems that overlaps with our analytical problem. If we exclude regional variance, then we cannot tell whether it is or is not a factor.

4. Figure 2.3 could, of course, be made to appear more complex. For instance, Charles Armstrong (2004: 44) notes some of the implications of the termination in 1975 of the North/South Vietnamese rivalry for the North/South Korean rivalry.

5. Assessing significant de-escalation versus termination is an inherently subjective exercise. In some rivalry situations, decision makers may be distracted by other concerns, thereby allowing the rivalry to "grow cold" or hibernate for a period of time before it heats up yet again. Even in cases in which the consensus is that a rivalry has ended, there are apt to be lingering doubts among some proportion of the elite and the population in general. Old habits die hard. Moreover, there is frequently some propensity to flirt with re-escalation, as is evident in the contemporary Sino-U.S. and Russo--U.S. cases. Significant de-escalation, therefore, might be defined as approximating termination (rivals no longer view each other as threatening competitors or enemies), but some doubts remain and the potential for the reactivation of tension persists.

Chapter 3

1. Technically, there was no Israeli state at the onset of the war in 1948. In addition, Nasser is considered the leader of the coup effort in 1952, but Neguib served as the official or figurehead leader of the Free Officers until he was removed in 1954.

2. This Greater Syria plan involved the political unification of Transjordan, Syria, Lebanon, and parts of Palestine under Hashemite rule.

3. Thomas Mayer (1983) also agrees that the Egyptian government's involvement in the Palestine problem was largely the result of its concern about playing the dominant role in the region vis-à-vis other Arab states.

4. For more details of these negotiations, see Caplan 1997 and Morris 1993.

5. The CIA had many contacts with the Free Officers Movement and approved the military coup (McNamara 2003; Sachar 1981).

6. The Israeli decision to raid Arab villages was an effort to influence the civilian population to put pressure on the Egyptian government to clamp down on the infiltrators. The Israelis shifted to military targets when the international community criticized these actions (Tal 1996: 65).

7. Shlaim (2001a: 118–120, 128) also documents high-level contacts between Nasser and Moshe Sharett, the Israeli prime minister, from October 1954 until January 1955. The talks covered specific problems dealing with the Egyptian capture of an Israeli ship, the *Bat Galim*, and the pending trial of Jewish spies who had been caught engaging in sabotage inside Egypt. These talks also covered broader issues dealing with Egypt's blockade of Israeli shipping in the Suez Canal and the Gulf of Aqaba, border skirmishes, the Palestinian refugees, and options for economic cooperation. Nasser even agreed to a high-level meeting with Sharett, but Sharett wavered and called off the talks. The Gaza raid in February 1955 ended these talks altogether.

8. As a result of Western pressure and heavy Egyptian casualties, Nasser stopped the Gaza raids. Since he knew that his army was considerably weaker than the Israeli Defense Forces, Nasser decided to de-escalate for fear that Israel's counterattacks

could jeopardize his control over Gaza and even the Egyptian army itself (Morris 1993: 352).

9. Morris (1993: 273) and Podeh (2004: 77) maintain that Egypt's seizure of an Israeli ship, the *Bat Galem*, in the Suez Canal in 1954 and the execution of two Jewish saboteurs in Cairo in January 1955 were critical events that soured Israel's relations with Egypt. Morris (1993: 273) indicates that the executions had an even greater impact when Prime Minister Sharett lobbied intensively for leniency to no avail. For more details on the espionage story, see Morris 1993 and Shlaim 2001a, with its discussion of the "Mishap."

10. Tal (1996: 64) argues that the Israelis perceived negative consequences from the Anglo-Egyptian agreement in 1954: a) the loss of a British buffer between Egypt and Israel; and b) enhanced military capabilities as the Egyptians took over former British military installations.

11. The arms deal resulted in a commitment by the Czech government to supply $320 million worth of arms to Egypt on an interest free basis in exchange for annual payment in Egyptian cotton. The quality of the war material was the highest yet received by Egypt and included 120 jet fighters, 50 bombers, 200 tanks, 150 artillery pieces, 2 destroyers, and 2 submarines. In addition, Soviet and Czech instructors were sent to train Egyptian soldiers on its new equipment (Sachar 1981: 91; Brecher 1972: 60).

12. The United States turned down Nasser's last request for arms in July 1955 (Morris 1993: 276).

13. In the late 1940s and early 1950s Britain and the United States had attempted to minimize the inflow of arms to the Middle East.

14. The Baghdad Pact was initiated by the United States and Great Britain in order to build a regional defense system that would be a bulwark against the Soviet Union. The pact involved cooperation for security and defense and a pledge of noninterference in the internal policies of member states.

15. One of the key goals for Nasser and the Free Officers was to eradicate the British military presence in the Suez Canal. They had secured that goal with the signing in 1954 of the Anglo-Egyptian Evacuation Agreement, which required British troops to be gone by June 1956. Once Nasser had secured Britain's withdrawal, he focused on destroying the Baghdad Pact, which threatened to prolong Britain's military presence in the region.

16. After the Baghdad Pact, Nasser solidified his ties with other Arabs by forming a military alliance with Syria and Saudi Arabia in March 1955. In April 1955, Nasser attended the Bandung conference for nonaligned states. Nasser, along with other leaders such as Zhou Enlai and Tito, advanced concerns outside the confines of the Cold War. After the conference, Nasser was more radical in his policies and more popular than any other Arab leader (McNamara 2003: 43).

17. This interpretation about the effect of the Czech-Egyptian arms deal has been challenged by recent historical evidence that suggests the Israeli officials were not

threatened by the arms deal but were willing to use the deal as an opportunity to justify joining the Anglo-French coalition in the Sinai War of 1956 (Golani 1955; Laron 2009). Even so, the arms deal intensified the intensity of the rivalry for both Egypt and Israel.

18. The Tripartite Agreement of 1949 stipulated that Britain, France, and the United States would refrain from selling arms to states in the Middle East.

19. Eisenhower believed that the conflict with Nasser over the Suez Canal could be resolved diplomatically, and he sent clear messages that he opposed war to resolve it. On principle, he believed that Nasser's action was not illegal and that he was in a position to ensure that the canal ran smoothly. He also believed that Egypt was the victim of aggression, and that Britain and France had violated the Tripartite Agreement of nonintervention (Neff 1981).

20. Despite the agreement about the Straits of Tiran, Nasser maintained a blockade of Israeli shipping through the Suez Canal.

21. In 1962 and again in 1965 Egypt faced serious balance-of-payments crises, which forced Nasser to rely heavily on foreign assistance provided largely from the Soviet Union. There were five major reasons for the economic crises: 1) the failure of the Egyptian cotton crop in 1961; 2) the costs of involvement in the Yemeni civil war; 3) suspension of U.S. food aid after 1965; 4) military spending; and 5) poor export performance (Waterbury 1983; Barnett 1992; Rivlin 1997).

22. Dayan later maintained that at least 80 percent and probably more of the clashes in the demilitarized zones were initiated by the Israelis (Slater 2002: 91).

23. The famous three no's (no peace, no negotiations, and no recognition of Israel) associated with the Khartoum Conference were not proposed by Nasser or any other Arab leader. Instead, they emerged from Ahmad al-Shuqayri, the head of the Palestine Liberation Organization (PLO), who wanted to prevent King Hussein of Jordan from pursuing a political solution for the West Bank. Although Nasser and the other Arab leaders were not in favor of this extreme stance, they accepted the position because of their unwillingness to look like they were surrendering to Israeli demands after their military defeat in the Six Day War (Shemesh 2008: 242). Yet, the three nays were flexible enough to leave open three options: 1) that Arab states might support a state of peace even if they did not sign a peace treaty; 2) that they might permit third-party negotiations to occur; and 3) that they might accept Israel, although they might not extend de jure recognition (Shlaim 2001a: 258).

24. UN Resolution 242 calls for the withdrawal of Israeli forces from territories it occupied in the aftermath of the Six Day War, and respect for the sovereignty among all states in the Middle East.

25. The purpose of the Jarring mission, headed by Dr. Gunnar Jarring, was to facilitate the implementation of UN Resolution 242. The mission ended in 1969 in the absence of any progress; it was revived again in 1971 for a short period, again with no success.

26. In the Six Day War of 1967, Egypt lost 80 percent of its military equipment; 10,000 of its soldiers and 1,500 of its officers were killed; 5,000 of its soldiers and 500 of its officers were captured; and 20,000 soldiers were wounded while Israel, in contrast, lost 779 soldiers on all military fronts (Hopwood 1991; Stone 2004). In its appeals to oil rich, Arab countries, Egypt explained that it had lost approximately $324 million annually from the closing of the Suez Canal and the loss of its oil fields (Feiler 2003: 121).

27. In 1967, approximately seventy thousand Egyptian soldiers were deployed in Yemen, constituting roughly half of Nasser's army. Schmidt Adams (1968) and Haddad (1973) maintain that Nasser was spending half a million dollars a day during the five-year war with Yemen—a figure that did not include the indirect costs of arms and supplies to the Yemenite army.

28. The Arab oil countries provided the aid for a myriad of reasons: a) reinforcement of Nasser's shift from an attempt to undermine conservative Arab regimes via a pan-Arab strategy; b) an attempt to refocus Nasser's ambitions toward Israel rather than dominating the Arab region; c) fear of more Soviet encroachment in the area if Egypt became more dependent on the Soviet Union for aid; d) an effort to balance Israeli military strength in the region; and e) securing greater political legitimacy for their regimes against Arab revolutionary states (Feiler 2003: 123–124).

29. The Israelis did not offer to give back the Gaza Strip.

30. Approximately five hundred aircraft shelters were built at twenty airfields around the country with underground operations rooms, pilots' quarters, and maintenance facilities. Sections of the highways between Cairo and the Suez Canal were widened and adjusted for emergency landing strips (Korn 1992: 92).

31. At this point, Nasser convinced the Soviets to supply Egypt with new surface-to-air missiles (SAM-3s) with the capacity to shoot down Israeli aircraft. With the aid of Soviet technicians who operated the missiles, the Egyptians were able to secure more antiaircraft protection from deep penetration raids by the Israelis. Eventually, the Egyptians moved the missiles gradually toward the Suez Canal in an effort to extend the antiaircraft shield to the east bank and provide cover for an eventual Egyptian crossing of the canal. The Israelis in turn attacked these sites without much success (Yaqub 2007: 41).

Chapter 4

1. This episode is referred to as the Corrective Revolution.

2. An opening for peace seemed possible in the early 1970s. For arguments against this view, see Gazit 1997.

3. Several other factors also played a role in the expulsion: a) Soviet reluctance to provide military hardware that Sadat requested; b) tension between Soviet and Egyp-

tian military officers; c) the Soviet Union's failure to meet commitments; and d) Soviet interference in Egyptian internal affairs (Feiler 2003: 131).

4. Isma'il would meet again with Kissinger in May 1973, but by March 1973 Sadat had decided to embark on a limited war with Israel (Beattie 2000: 131; Seale 2000: 47).

5. The United States also built an early-warning station for Egypt and manned it with U.S. soldiers.

6. In 1976, Egypt was late in paying $800 million already in arrears (Karawan 2005: 329).

7. Of this amount, almost 1 billion Egyptian pounds were allocated for food commodities. This is compared with 690 million Egyptian pounds for subsidies in 1978 (480 million pounds for food), 435 million in 1974, and 255 million in 1973 (Rivlin and Even 2004: 42).

8. Barnett (1992: 130) maintains that there were four reasons for the increased national security outlays: 1) Egypt needed to replenish stocks as Israel was doing; 2) Egypt feared threats from Soviets and other regional actors; 3) the supplies of U.S. arms were more expensive; and 4) a strong defense would attract foreign capital and assistance.

Chapter 5

1. The Israeli-Syrian analysis is restricted to the years 1948–2000 to make it more complementary to the Egyptian-Israeli case that preceded it.

2. The British and the French negotiated a deal when they had mandate control over Israel/Palestine and Syria. The British handed over the south Golan area to Syria; the French gave up a slice of territory along Lake Tiberias to Palestine (under British mandate), thus enclosing Lake Tiberias within Palestine territory.

3. This interpretation is based on Shlaim's (2001a: 30) account.

4. The numbers are found in Ma'oz's (1995: 89–93) discussion.

5. Dayan later admitted that at least 80 percent and probably more of the clashes in the demilitarized zones were initiated by the Israelis (Slater 2002: 91).

6. The constraints are discussed in Shlaim 2001a: 285–289.

7. Meir had opportunities to reevaluate Israeli relations with the Arabs, but she ignored them. For instance, Meir turned down a UN suggestion that Israel return Sinai to Egypt in return for peace—a decision that Israel would make at the end of the 1970s. She also turned down Sadat's offer for an interim settlement (Shlaim 2001a: 324).

8. Asad believed that military parity produced peace, while military imbalances encouraged war as the stronger party would always be tempted to secure its aims by force. Hence, Syria and the rest of the Arab countries should build enough military power to deter Israeli military advances in the future (Seale and Butler 1996: 31–32).

9. Syria eventually deployed SAM sites in Zahla (central Lebanon) after clashing with Israeli aircraft over the Biqa Valley in April 1981. The missile crisis occurred after Maronite forces attempted to secure Zahla, which had been held by the Syrians since 1976. The Maronites' efforts jeopardized Syria's access to the Beirut-Damascus Highway, a key artery linking Syria to Lebanon. The Syrians used their air force to neutralize the Maronites, thus provoking military air strikes from Israel in Lebanese air space. The crisis eventually ended when the United States negotiated a ceasefire (Seale 1988: 371).

10. Ma'oz (1995: 204) documents that Asad's comments were made in public after personal meetings with former president Jimmy Carter and conversations with Senator Arlen Specter.

11. The Soviet action reflected a new global strategy that eschewed Soviet military involvement in Third World conflicts in favor of supporting regional peace settlements.

12. Consequently, Asad chose to join the U.S.-led international coalition against Iraq in the Persian Gulf War of 1991 with the understanding that the United States would pursue a peace strategy after the war.

13. Radical Arab states and the PLO lost regional and international prestige as a result of their support for Hussein during the Persian Gulf War of 1991, smoothing the way for the moderates to search for a political settlement.

14. Asad formally announced his "full peace for full withdrawal" position in an interview with Patrick Seale that was published in both Arab and Western newspaper outlets (al-Asad and Seale 1993).

15. Full details about these agreements are provided in Slater 2002.

16. Later, the Israeli government would disavow Rabin's commitment, but later still in 2000, Prime Minister Ehud Barak of Israel acknowledged that Rabin had indeed made the agreement in July 1994 and that he considered it binding on the Israeli government. The full details of this story can be found in al-Moualem 1997; Cobban 1999; Hinnebusch 1996; and Seale 2000.

17. According to Hinnebusch (1996: 55), Syria's negotiating strategy was always based on "detailed, comprehensive and iron-clad agreements" with specific timetables for Israeli withdrawals. Moreover, Israel's subsequent interpretations of the Oslo Accords at the PLO's expense only reinforced Asad's determination to avoid a similar trap.

18. Israel and Jordan agreed on a peace agenda one day after the signing of the Oslo Accords in Washington, D.C. At that point, the peace negotiations moved through a series of phases, culminating in a peace treaty that was signed on October 26, 1994 (Seale 2000: 68).

19. A public opinion survey in Israel during January 1994 revealed that most respondents were unwilling to give back the Golan in return for full peace and security arrangements (Ma'oz 1995: 242).

20. Operation Grapes of Wrath was targeted primarily at Syria. Israeli military planners decided to bomb southern Lebanon with the idea of routing civilians from

southern to northern Lebanon so that Israeli troops could strike more heavily against Hezbollah forces in the south. In addition, the operation was designed to put pressure on the Syrians to rein in Hezbollah's actions (Shlaim 2001a: 560).

21. This quote was found in Cobban 1999: 180.

22. Rabin also argued that Israelis could not digest progress in more than one track involving the Syrians and the Palestinians (Cobban 1999: 186).

23. Asad was particularly adverse to pursuing "Sadat-like acts of diplomacy" since he perceived that Sadat had betrayed Syria and the Arab cause during the Camp David negotiations in 1979 (Kessler Neff 2000: 79).

24. Interfactional competition for power and influence on foreign policy, consequently, tended to translate into free and lively debates, rather than real planning and implementation within both the PLO and Fatah (Groth 1994). Although urgent matters such as the PLO's evacuation from Beirut in 1982 cut across factional lines and made the PLO able to respond quickly to short-term critical problems, such problems involved primarily tactical responses. Planning strategies for the long term were confined to the slower and more politicized bodies of the PLO—and, therefore, were unlikely to be forthcoming.

25. Fatah militants argued that Arafat's moves were illegitimate because they occurred outside the routine PLO decision-making bodies, undermined the strategy of military struggle, and abandoned the goals of an independent Palestinian state (see Sahliyeh 1986: 142–153).

26. In fact, the Palestinian National Council extended Arafat's authority by allowing Arafat and his colleagues within the PLO Executive Committee to decide on the issues of a political settlement in coordination with Jordan and Egypt. Arafat's freedom of maneuver was also enhanced by the appointment of leading moderate figures to key positions within the PLO (Sahliyeh 1986: 180–202).

27. Later, eighty-four countries extended full recognition to the new state of Palestine and some twenty others offered qualified recognition. In January 1989, the PLO gained the right to address the UN Security Council on an equal footing with other member states (Sayigh 1997: 624).

28. Arafat's support for the Iraqi invasion, in turn, was due at least in part to his efforts to replace the lost Soviet support for the PLO (Rubin 1994: 155).

29. The accords provided for the establishment of a self-governing authority by the PLO in Gaza and Jericho. Palestinian authority would extend eventually to the remaining Palestinian population centers of the West Bank in a second phase, which would coincide with general elections to form a governing council. After these interim arrangements, further negotiations were expected to determine the final status of Jerusalem, Israeli settlements in the occupied territories, the fate of Palestinian refugees, and other matters as part of a permanent settlement of the Israeli-Palestinian conflict.

30. Israeli analysts also argue that the large influx of new Soviet immigrants had a significant impact on the outcome of the election of 1992. Surveys indicated that Soviet immigrants were less interested in ideology and foreign policy than the issues of

absorption (that is, employment and housing). Labor and other left-wing parties were able to link foreign policy and absorption issues in a way that was more attractive than the Likud Party. Labor argued that if the government spent fewer resources on building projects in the territories, then there would be more monies available for dealing with the problems of unemployment and inadequate housing. This issue increased the trend among the immigrants to support territorial concessions and the Labor Party in the 1992 election (Fein 1995: 168–173). It was readily acknowledged (Reich et al. 1995: 142) that the Soviet vote was not so much a vote in favor of Labor but a protest against a Likud government that had failed to respond adequately to the needs of the new immigrants. It was also a demand for the next government to reorient its domestic spending priorities.

31. Rabin initially rejected the PLO as a negotiating partner because of its insistence on a state encompassing all of the West Bank and Gaza and its claim to represent all of the Palestinians in the diaspora with dreams of return that Israeli decision makers could not accommodate. Rabin believed (Makovsky 1996: 110) that a debilitated PLO would be forced to acquiesce to a deal made by local Palestinians in order to ease the occupation.

Chapter 6

Sumit Ganguly thanks Manjeet Pardesi for his research assistance.

1. A summit had taken place in Agra, outside New Delhi, between President Pervez Musharraf and Prime Minister Atal Behari Vajpayee in July 2001. However, this attempt at de-escalation achieved next to nothing and proved to be mostly an exercise in grandstanding on Musharraf's part. Also, it is believed that L. K. Advani, India's hard-line deputy prime minister and minister of home affairs, played a role in undermining the talks. On this issue, see Jacob 2010 and Noorani 2005.

2. On the significance of credible commitments in strife-ridden polyethnic societies, see Fearon 1998.

3. In recent years a controversy has arisen over the question of the contiguity of Kashmir with India. The controversy stemmed from the allegations of a British historian, Alastair Lamb, about Mountbatten's putative pressure on Sir Cyril Radcliffe, the barrister who had been given the task of dividing the empire. Lamb alleges that Radcliffe, under pressure from Mountbatten, drew the boundaries to favor Kashmir's accession to India. For Lamb's allegation, see Lamb 1992; for a careful and spirited rebuttal, see Ilahi 2003.

4. There was, however, one brief moment of significant cooperation. This was the agreement to share the Indus River waters signed in 1950 under World Bank auspices.

5. For a critical account of the Eisenhower administration's decision to forge a military relationship with Pakistan, see McMahon 1994.

6. The notion that Pakistan, the putative homeland of the Muslims of South Asia, could be a bulwark against Communist expansion was a colonial construct. Sir Francis Tuker, a senior British general, propagated this idea at the time of independence and partition (Tuker 1950).

7. A number of explanations can be adduced for India's unwillingness to increase defense spending. At one level, it would compromise India's ideational commitment to nonalignment. At another, Nehru was acutely concerned about the opportunity costs of defense spending. On Nehru's views, see Cohen 1990; for a discussion of India's defense spending, see Thomas 1986; for a sense of the anxieties, see Dutt 1977. Dutt was a senior foreign policy official and eventually retired as India's foreign secretary, the highest-ranking bureaucrat in the Ministry of External Affairs.

8. For a dispassionate account, see Hoffmann 1990; the oft-quoted, but utterly polemical, version of the war can be found in Maxwell 1970.

9. One of the best accounts of this military debacle remains Palit 1991.

10. Substantial military assistance from the United States was not forthcoming because the United States was most reluctant to alienate Pakistan. The United States was acutely dependent on Pakistan for access to key military bases. These bases served as listening posts on the then Soviet Union and were also used for U-2 flights.

11. This was not a true "security dilemma" for Pakistan; it was not as if India's military modernization program posed any threat to Pakistan's security. Nevertheless, Pakistani rhetoric echoed the concerns of a security dilemma. On the concept of "security dilemma," see Herz 1950.

12. For details pertaining to this war, see Jackson 1975 and Sisson and Rose 1991.

13. Curiously enough, Pakistani decision makers, on the basis of the most tenuous evidence continued to believe that China would open a second front and thereby tied down Indian troops along the Himalayan border. Recently declassified documents from the U.S. State Department reveal the depth of the delusion of Pakistani decision makers, most notably Yahya Khan. For details, see Smith 2005.

14. His quest was far from chimerical. The initial supply of U.S. assistance was for some $3.2 billion for five years. The subsequent tranche was for $4.02 billion for six years.

15. For a discussion of the concept of reciprocity in international relations, see Kydd 2005.

16. It was also inhibited from undertaking a military misadventure because of the Soviet presence in Afghanistan.

17. It needs to be made clear that Indian and Pakistani interlocutors hold markedly divergent views about the Shimla Accord. For the Indians, the accord enshrined the principle of bilateralism. For the Pakistanis, it did not. It merely affirmed that bilateral negotiations were one possible pathway for the settlement of the dispute. For a strident and forceful exposition of the Pakistani position, see Sattar 1995.

18. It is pertinent to mention that the Soviet collapse was one of the most dramatic shocks for India. India had developed a robust arms transfer relationship with the

Soviet Union since 1971. More to the point, India had come to depend on the Soviet Union for a tacit security guarantee against the People's Republic of China. While the Soviet collapse did lead to a dramatic reorientation in Indian foreign policy it had little or no effect on the Indo-Pakistani rivalry. On the evolution of the Indo-Soviet relationship, see Horn 1982; for the reorientation in Indian foreign policy at the end of the Cold War, see Mohan 2004.

19. Throughout the 1990s, the Indian state, through a combination of widespread repression and limited political accommodation managed to contain the insurgency and restore order, if not law, in Kashmir. This counterinsurgency strategy, which had worked well elsewhere in India, continues to be applied with some success in Kashmir.

20. The literature on Kargil, though almost exclusively from the victorious Indian side, is voluminous. See, for example, Krishna and Chari 2001; Swami 1999; Bajpai, Karim, and Mattoo 2001 and Singh 2001. For a set of U.S., Indian, and Pakistani perspectives on the Kargil war, albeit of intellectually uneven perspectives, see Lavoy 2009.

21. Details pertaining to the dialogue remain scanty. For one account, see Coll 2009.

22. In 2005, the Indian prime minister, Manmohan Singh, while willing to improve relations with Pakistan, categorically ruled out any prospect of territorial compromise. See the prime minister's statement in the Lok Sabha (Parliament), April 20, 2005 at http://www.pm.nic.in.

23. See "The ISI Controlled 26/11 Attacks from Beginning to End," *Indian Express*, July 17, 2010, available at http://www.indianexpress.com/news/the-isi-controlled-coordinated-26-11-attacks-from-beginning-to-end/647849/0.

24. The desire to hold the Indian state together is the reverse side of the coin vis-à-vis the initial emphasis on a secular state.

Chapter 7

1. Of course, we do not mean to suggest that earlier chapters have not raised the issue of rivalry interdependency—only that they have not stressed it to the extent that we propose to do in this chapter.

2. Dittmer 2011 surveys the long history of the Russo-Chinese relationship.

3. See, for instance, Thompson's (2003) analysis of rivalry interdependence prior to the outbreak of World War I..

4. The Cold War rivalry requires its own book.

5. General sources on the Sino-U.S. rivalry include Shambaugh 1994, Ross 1995, Garver 1997, and Scott 2007.

6. Zhang Baijia (2001) notes that the United States as the world's leading imperial power also made ideological conflict between China and the United States more probable. On this theme, see also Harding 1992: 25–26.

7. Tyler (1999: 9) describes Taiwan as "an unsinkable aircraft carrier to contain the Chinese 'menace.'"

8. Harding (1992: 33) argues that at least some Chinese decision makers considered altering their stance toward the superpowers in the early 1960s in the face of increasing military threat from the Soviet Union and major economic problems at home linked to the earlier attempted Great Leap Forward project. Foot (1995: 97–98) provides a brief overview of U.S. contemplation of creating an opening to China in 1962. This effort was derailed by, among other things, the Sino-Indian War and the increased emphasis given to U.S.-Soviet relations after the Cuban Missile Crisis.

9. Dittmer (2004: 208), for one, argues that the invasion of Czechoslovakia in 1968 led to the border clashes initiated by the Chinese in 1969.

10. Between 1969 and 1971 (and continuing into the early 1980s), the Soviet Union greatly increased the number of troops stationed along its border with China (Scott 2007: 80).

11. Domestic turmoil was a problem for the Chinese, but it also made them less threatening to the United States (Schulzinger 2001: 261). The U.S. defense budget planning process of 1969 reduced its worst-case scenario from 2.5 to 1.5 wars. The older conception that the Soviet Union and China might initiate joint attacks was finally abandoned and China's own threat to the United States was viewed as diminished (Ross 1995: 23).

12. Somewhat in parallel, the Chinese were worried that détente between the United States and the Soviet Union might encourage greater Soviet pressure on China (Ross 1995: 25). Garver (1982) stresses that the Chinese were primarily interested in ensuring that the Soviet Union and the United States remained rivals as opposed to the ongoing trend of increased superpower cooperation and agreement that China was a major mutual problem.

13. Nixon is reputed to have been irritated by Vice President Agnew's criticism of the overtures to China and said that he was missing "the big picture in this whole Chinese operation, which is, of course, the Russian game. We're using the Chinese thaw to get the Russians shook" (quoted in Schaller 2001: 373).

14. Decision makers in the United States apparently also sought greater cooperation from Japan with the opening to China (Ross 2001; Schaller 2001).

15. Li Jie (2001: 313) starts the signaling clock in November 1968 with China's proposal to the outgoing Johnson administration to resume ambassadorial talks in Warsaw. Gong Li (2001: 332) contends that Mao had read Nixon's article in *Foreign Affairs* prior to the election; the article suggested the need to reconsider U.S. isolation and containment of China and as a consequence, Mao thought an election victory by Nixon would work in China's diplomatic favor.

16. Demonstrating some appreciation for evolutionary conceptualization in rivalry termination processes, Mao is said to have commented that monkeys required some time to evolve into humans. He compared the U.S. position after one of Kissinger's

preliminary talks with Zhou to that of an ape with a tail that was in the process of becoming shorter (Schaller 2002: 175).

17. General sources on the Sino-Soviet rivalry include Dittmer 1992, Zhao 1996, and Rozman 2010.

18. Nelson (1989: 139) says the downgrading came later.

19. This seems to remain true, although probably less so, with the resumption of the rivalries between China and the Unites States and China and Russia.

20. General sources on the Sino-Vietnamese rivalry include Ross 1988, Hood 1992, Amer 1999, Westad 2006, Womack 2006 and 2011, and Rozman 2010.

21. The Chinese may also have been attempting to reassure their Thai allies as well as warning the Soviet Union how intense their unhappiness with Vietnamese behavior was (see, for instance, Stuart-Fox 2003: 201).

22. Information on the Thai-Vietnamese rivalry is scarce. See Sar Desai 1994, Duiker 1995, and Wyatt 2003.

23. On the Chinese-Taiwanese rivalry, see Gu 1995, Myers and Zhang 2006, Kastner 2007, Wachman 2007, R. Sutter 2008, and Zhao and Liu 2010.

24. See, for instance, Huang, Kim, and Wu 1995: 36.

25. Useful sources on the rivalry between the two Koreas include Kim 1994 and 2011, Gu 1995, Oberdorfer 2001, Heo and Roehrig 2010, and Snyder 2010.

26. Kim Jong Un succeeded Kim Jong Il in December 2011. No doubt, it will take the new ruler some time to consolidate his leadership position. Moreover, there is no reason to assume that such a leadership consolidation must be successful. But it is reasonable to assume that a strong leadership position would be essential to bringing about major revisions in North Korean foreign policy strategies.

27. Gu (1995: 189) characterizes the military conflict periodicity of the Korean rivalry as hot war (1948–1953), standoff/cold war (1953–1965), low-intensity hot war (1965–1971), standoff/cold war (1972–1992), and disengagement from 1992 onward. Yet it would be difficult to argue that the degree of disengagement has been constant since 1992. See Hoare and Pares 1999: 231–40; Oberdorfer 2001; Mauryin 2002; Nanto 2003; and Kim 2011. Cumings (2004: 56), parenthetically, also argues that western observers are more familiar with northern incursions into the south and less familiar than they should be about southern incursions into the north.

28. Porter (1993: 207) argues that Vietnam was looking for a Cambodian "exit strategy" as early as 1984.

29. The irony here is that domestic political reforms were initiated for the purpose of ensuring the security of the domestic regime (Cliff 1998) yet may have increased insecurity at the external national level. As Kahn (2004) suggests, however, electoral politics in Taiwan need not lead to greater interstate hostility.

30. Domestic economic conditions, short of crisis, can also interact with domestic politics toward external rivalry processes. For example, Taiwanese businessmen responded to the economic recession in the late 1990s with stronger demands for a re-

laxation of restrictions on investments and trade with the Chinese mainland (S. Goldstein 2002; Sheng 2002).

31. See Cabestan (2002) for some speculation on scenarios that might lead to greater integration without formal reunification.

32. Oberdorfer (2001: 273) notes that the suspension of the Team Spirit exercise in 1992 was considered the "most tangible evidence of an improved relationship" between the two Koreas. Thus, the loss of this tangible evidence would also seem to indicate the end of an improved relationship.

33. China and Vietnam still do not agree on the resolution of the Spratly Islands dispute and occasional temptations to take more hard-line stances toward China still surface (see, for example, Ninh 1998).

34. Perhaps the most significant obstacle to improved inter-Korean relations is the conflict relationship between the United States and North Korea. It is difficult for inter-Korean relations to remain cooperative as U.S.–North Korean relations deteriorate. Seoul's location and its consequent vulnerability to an attack is simply too convenient a bargaining chip for North Korea not to use it when its other options are quite limited.

35. Third-party encouragement for de-escalation seems just as likely to be rebuffed if local decision makers do not have local incentives to pursue rivalry de-escalation for their own reasons. See, for example, Wishnick's (2004: 121–122) discussion of Russia's encouragement of North Korea to cooperate with South Korea in 2001.

36. Several authors have noted in this respect that Chinese-Vietnamese territorial disputes along their mutual border reflect more general tensions between the two states than they do the intensity with which decision makers value the specific territorial boundaries (see Sutter 1993: 45; Wang 2003: 401).

37. That is not quite the same thing as saying that all personnel must remain in power. A Vietnamese foreign minister was sacrificed to mollify the Chinese.

38. Christensen (2004: 66) argues that the issue of Taiwan is linked in part to the fear of a national territorial disintegration on the mainland. An independent Taiwan could encourage other areas to attempt to break away.

39. See, for instance, the discussions in Chun 1997 and Noland 1997. It is not inconceivable, however, that the North Korean regime might be overthrown from within, thereby removing one major obstacle to rivalry termination. See, for instance, the speculative discussion (Brooke 2004: A3) pertaining to some 10 percent of the North Korean army's generals who have allegedly defected in recent years.

40. At times South Koreans have worried that North Korea and the United States might make a separate peace that ignored South Korean preferences (Moon 1997). The strong, if intermittent, role of the United States in North–South Korean relations must be considered a complication for de-escalation purposes. Cha (2004) contends that the United States is not a permanent impediment to better inter-Korean relations.

But, in the long run, the strong U.S. role in the Korean peninsula may also demonstrate the reverse effect of usual expectations about third-party pressures; that is, that third-party pressures may be as likely to impede de-escalation as they are to facilitate it. If cooperation and accommodation must be achieved simultaneously in two dyads, as opposed to one (as in South and North Korea and the United States and North Korea), then the odds against successful de-escalation are greater (see Armstrong 2004: 51). For that matter, it is unlikely that there would have been a North–South Korean rivalry without U.S. intervention early on. The same could be said of the rivalry between China and Taiwan.

41. We also need to know more about triangular rivalry relations. See, for starting points, Dittmer 1981 and Goldstein and Freeman 1990.

Chapter 8

1. At the same time, we do not regard the coerced/subordinated category of terminations to be either noncomparable or beyond the modeling pale. They frequently involve shocks, especially war defeats, and certainly encompass expectancy revisions about who is competitive with whom. The difference is that the change in circumstances is so stark that there tends not to be anything more to negotiate about. Often, the victor can dictate the resolution of any outstanding disputes.

2. In this volume, we engage in qualitative case studies of instances with varying outcomes. Rasler (2001b) demonstrates, however, that the model can also be tested quantitatively.

3. For instance, Sadat, Gorbachev, and Rabin have all been depicted as absolutely necessary to the de-escalations of the Egyptian-Israeli, U.S.-Soviet, and Israeli-Palestinian rivalries, respectively. The problem is that it is impossible to rerun the processes in the real world with different people to see whether the outcome was indeed dependent on the personalities involved. This might be a puzzle that could be explored, however inconclusively, with experimentation.

4. We regard the first Chinese-U.S. rivalry as Eurasian in nature because its focus was placed on East and Southeast Asian issues. The ongoing, second Chinese-U.S. rivalry has been Eurasian in nature but it has some potential for breaking out of that mould eventually.

Appendix

1. One model that we do not include is the one developed by Long and Brecke (2003) to explain postwar reconciliation. To our minds, their approach is not designed to deal with interstate rivalry termination processes per se. Reconciliation is some-

thing that may follow rivalry termination, but there may be a considerable lag. The stress on reconciliation works better for civil wars, as they note, than it does for interstate wars. Symbolism about forgiveness is more likely to be needed after internal combat than it is after fighting between two states. Unfortunately, we became aware of Mani 2011 too late to include it.

2. One of the ironies of this exercise is that our own model preceded many of these arguments. Hence, we did not really construct a synthesis based on earlier works but, rather, have developed a model that appears to serve as a selective synthesis of various other arguments that have been put forward on kindred topics.

3. See as well the oyster shell model imagery discussed in Rasler and Thompson 1989, 16–17.

4. It should be noted that Legro's argument is not focused on rivalries but, rather, on major power grand strategies. We see no reason, however, that the argument can apply only to grand strategies or, for that matter, major powers.

5. Maoz and Mor (2002) focus primarily on the Egyptian-Israeli and Israeli-Syrian rivalries.

6. This observation is not part of Maoz and Mor's theory but helps to explain how some rivalries have experienced a series of war iterations without necessarily influencing perceptions of the status quo or of capabilities.

7. Maoz's more recent work has departed significantly from an insistence on dyadic structures. See Maoz 2011 for an emphasis on situating states within systemic networks.

8. The main exception is that decision makers can manipulate the salience of geopolitical objectives.

9. See Leng 1983 and Colaresi, Rasler, and Thompson 2007 on the problem of serial crises in rivalry relations.

10. The classic case is Britain's de-escalation of its French, U.S., and Russian rivalries to better cope with the German threat prior to World War I.

11. Tony Armstrong's (1993) case focus is placed primarily on East and West Germany.

12. These preference schedules can be constructed at the individual level and it is not unusual to find some considerable variance exhibited within any given decision-making group. To aggregate at the national level, therefore, involves considerable reduction.

13. Pervin's treatment is fairly brief and, given more space, it is likely that more than one Arab preference schedule would have been constructed. One schedule cannot encompass equally well Jordanian, Syrian, and PLO preferences.

14. In certain chapters we also emphasized the compounded effects of multiple shocks as manifested in the ending of the Cold War, the intifada, and the first Gulf War.

15. The inflexibility of decision makers is sufficiently frequent to cast doubt on assuming that all decision makers seek to remain in power above all else.

16. We agree with Pervin on the variable impact of third-party pressures as another part of the agency process.

17. Goertz and Diehl (1997) and Diehl and Goertz (2000) have been most responsible for injecting the concept of shock into rivalry analyses. While we acknowledge the significance of shocks, we have a much different take on what actually qualifies as a shock. See, for instance, the critique in Colaresi 2001. The basic points are that some types of shocks are much more important than others and that, even so, the effects of shocks are difficult to predict in isolation. Sometimes they open up windows for de-escalation, sometimes they intensify the rivalry, and sometimes they seem to have little effect.

REFERENCES

Abu-Amr, Ziad. 1993. "Hamas: A Historical and Political Background." *Journal of Palestine Studies* 22, 4 (Summer): 5–19.

Adler, Emanuel. 1991. "Cognitive Evolution: A Dynamic Approach for the Study of International Relations and Their Progress." In Emanuel Adler and Beverly Crawford, eds., *Progress in Postwar International Relations*. New York: Columbia University Press.

Ahmed, Samina. 1999. "Pakistan's Nuclear Weapons Program: Turning Points and Nuclear Choices." *International Security* 23, 4: 148–177.

al-Asad, Hafiz and Patrick Seale. 1993. "Interview with Syrian President Hafiz al-Asad." *Journal of Palestine Studies* 22, 4 (Summer): 111–121.

Allison, Graham T. 1971. *Essence of Decision: Explaining the Cuban Missile Crisis*. Boston: Little, Brown.

Al-Moualem, Walid. 1997. "Fresh Light on the Syrian-Peace Negotiations: An Interview with Ambassador Walid al-Moualem." *Journal of Palestine Studies* 26, 2 (Winter): 81–94.

Amer, Ramses. 1999. "Sino-Vietnamese Relations: Past, Present and Future." In Carlyle A. Thayer and Ramses Amer, eds., *Vietnamese Foreign Policy in Transition*. New York: St. Martin's Press.

Anderson, Jennifer. 1997. *The Limits of Sino-Russian Strategic Partnership*. Adelphi Paper 315. New York: Oxford University Press.

Arian, Asher and Michal Shamir. 1995. "Two Reversals: Why 1992 Was Not 1977." In Asher Arian and Michal Shamir, eds., *The Elections in Israel 1992*. Albany: State University of New York Press.

Arif, Khalid Mahmud. 1995. *Working with Zia: Pakistan's Power Politics, 1977–1988*. Karachi: Oxford University Press.

Armstrong, Charles K. 1998. "A Socialism of Our Style: North Korean Ideology in a Post-Communist Era." In Samuel S. Kim, ed., *North Korean Foreign Relations in the Post-Cold War Era*. Hong Kong: Oxford University Press.

———. 2004. "Inter-Korean Relations: A North Korean Perspective." In Samuel S. Kim, ed., *Inter-Korean Relations: Problems and Prospects*. New York: Palgrave-Macmillan.

Armstrong, Tony. 1993. *Breaking the Ice: Rapprochement between East and West Germany, the United States and China, and Israel and Egypt.* Washington, D.C.: U.S. Institute of Peace.

Aronoff, Myron J. and Yael S. Aronoff. 1998. "Domestic Determinants of Israeli Foreign Policy: The Peace Process from the Declaration of Principles to the Olso II Interim Agreement." In Robert O. Freedman, ed., *The Middle East and the Peace Process: The Impact of the Oslo Accords.* Gainesville: University of Florida Press.

Ashraf, Tariq M. 2004. "Doctrinal Reawakening of the Indian Armed Forces." *Military Review* 84, 6 (November-December): 53–62.

Avi-Hai, A. 1974. *Ben-Gurion, State Builder: Principles and Pragmatism, 1948–1963.* New York: Halstead Press.

Bajpai, Kanti P. , et al. 1995. *Brasstacks and Beyond: Perception and Management of Crisis in South Asia.* New Delhi: Manohar.

Bajpai, Kanti, Asfir Karim, and Amitabh Mattoo, eds. 2001. *Kargil and After.* New Delhi: Har-Anand.

Bar-Joseph, Uri. 2006. "Last Chance to Avoid War: Sadat's Peace Initiative of 1973 and Its Failure." *Journal of Contemporary History* 41, 3: 545–556.

Barnett, Michael. 1992. *Confronting the Costs of War: Military Power, State, and Society in Egypt and Israel.* Princeton, N.J.: Princeton University Press.

Barnett, Michael, and Jack Levy. 1991. "Domestic Sources of Alliances and Alignments: The Case of Egypt, 1962–1973." *International Organization* 45, 3: 369–395.

Bar-On, Mordechai. 1994. *The Gates of Gaza: Israel's Road to Suez and Back.* New York: St. Martin's Griffin.

——. 1996. *In Pursuit of Peace: A History of the Israeli Peace Movement.* Washington, D.C.: United States Institute of Peace.

Bar-Siman-Tov, Yaacov, 1999. Israel and the Intifada: Adaptation and Learning. Paper prepared for the Conference "Historical Roots of the Middle East Post-1945: Lessons for the Future?," Stiftung Bruno Kreisky Archiv, Vienna, 10–11 October 1999.

Bar-Tal, Daniel. 2000. "From Intractable Conflict through Conflict Resolution to Reconciliation: Psychological Analysis." *Political Psychology* 21, 2: 351–365.

Barzilai, Gad. 1996. *Wars, Internal Conflicts, and Political Order: A Jewish Democracy in the Middle East.* Albany: State University of New York Press.

Baumgartner, Frank R., and Bryan D. Jones. 1993. *Agendas and Instability in American Politics.* Chicago: University of Chicago Press.

Beattie, Kirk. 2000. *Egypt During the Sadat Years.* New York: Palgrave.

Bennett, D. Scott. 1997. "Democracy, Regime Change and Rivalry Termination." *International Interactions* 22, 4: 369–397.

——. 2004. "Patterns of Conflict Management and Resolution in Enduring Rivalries." In Zeev Maoz, Alex Mintz, T. Clifton Morgan, Glenn Palmer, and Richard J. Stoll, eds., *Multiple Paths to Knowledge in International Relations.* Lanham, Md.: Lexington Books.

Bennett, D. Scott, and Timothy Nordstrom. 2004. "Foreign Policy Substitutability and Internal Economic Problems in Enduring Rivalries." *Journal of Conflict Resolution* 44: 33–61.

Ben-Yehuda, Hemda. 1997. "Attitude Change and Policy Transformation: Yitzhak Rabin and the Palestinian Issue, 1967–95." In Efraim Karsh, ed., *From Rabin to Netanyahu: Israel's Troubled Agenda*. London: Frank Cass.

Ben-Zvi, Abraham. 1993. *The United States and Israel: The Limits of the Special Relationship*. New York: Columbia University Press.

Bercovitch, Jacob. 1995. *Resolving International Conflict: The Theory and Practice of Mediation*. Boulder, Colo.: Lynne Rienner.

Bhargava, G. S. 1983. *South Asian Security after Afghanistan*. Lexington, Mass.: Lexington Books.

Bhutto, Zulfikar Ali. 1969. *The Myth of Independence*. Oxford: Oxford University Press.

Bogumil, David D. 2002. "The Dynamics of Reciprocity and the Attributional Assessment of Perception and Behavior." *Social Behavior and Personality* 30, 2: 169–193.

Boulding, Kenneth E. 1970. *A Primer on Social Dynamics*. New York: Free Press.

Brines, Russell. 1968. *The Indo-Pakistani Conflict*. New York: Pall Mall.

Brooke, James. 2004. "Japanese Official Warns of Fissures in North Korea." *New York Times*, November 22.

Brown, Judith M. 2003. *Nehru: A Political Life*. New Haven, Conn.: Yale University Press.

Bueno de Mesquita, Bruce, Randolph M. Siverson, and Gary Woller. 1992. "War and the Fate of Regimes: A Comparative Analysis." *American Political Science Review* 86, 3: 638–646.

Bunce, Valerie. 1981. *Do New Leaders Make a Difference? Executive Succession and Public Policy under Capitalism and Socialism*. Princeton, N.J.: Princeton University Press.

Burton, Michael and John Higley. 1998. "Political Crises and Elite Settlements." In Mattei Dogan and John Higley, eds., *Elites, Crises,and the Origins of Regimes*. Lanham, Md.: Rowman & Littlefield.

Cabestan, Jean-Pierre. 2002. "Integration without Reunification." *Cambridge Review of International Affairs* 15: 95–104.

Caplan, Neil. 1997. *Futile Diplomacy: The United Nations, the Great Powers and Middle East Peacekeeping 1948–1954*. London: Frank Cass.

Cha, Victor D. 2003. "South Korea: Anchored or Adrift?" In Richard J. Ellings and Aaron L. Friedberg, eds., *Strategic Asia 2003–04: Fragility and Crisis*. Seattle, Wash.: National Bureau of Asian Research.

———. 2004. "The U.S. Role in Inter-Korean Relations: Container, Facilitator, or Impeder?" In Samuel S. Kim, ed., *Inter-Korean Relations: Problems and Prospects*. New York: Palgrave-Macmillan.

Chari, P. R., Pervaiz Iqbal Cheema, and Stephen P. Cohen. 2007. *Four Crises and Peace Process: American Engagement in South Asia*. Washington, D.C.: Brookings Institution Press.

Checkel, Jeffrey T. 1997. *Ideas and International Political Change: Soviet and Russian Behavior and the End of the Cold War.* New Haven, Conn.: Yale University Press.

Cheema, Pervaiz Iqbal, and P. R. Chari. 2001. *The Simla Agreement 1972: Its Wasted Promise.* New Delhi: Manohar.

Chiozza, Giacomo and H. E. Goemans. 2004. "International Conflict and the Tenure of Leaders: Is War Still Ex Post Inefficient?" *American Journal of Political Science* 48, 3: 604–619.

Christensen, Thomas J. 2004. "Chinese Realpolitik." In Guoli Li, ed., *Chinese Foreign Policy in Transition.* New York: Aldine de Gruyter.

Chun, Hong-Tack. 1997. "Economic Conditions in North Korea and Prospects for Economic Reform." In Thomas H. Henrickson and Jongryn Mo, eds., *North Korea after Kim Il Sung: Continuity or Change?* Stanford, Calif.: Hoover Institution Press.

Cliff, Roger. 1998. "Taiwan: In the Dragon's Shadow." In Muthiah Alagappa, ed., *Asian Security Practice: Material and Ideational Influences.* Stanford, Calif.: Stanford University Press.

Cobban, Helena. 1999. *The Israeli-Syrian Peace Talks: 1991–96 and Beyond.* Washington, D.C.: United States Institute of Peace.

Cohen, Stephen P. 1990. *The Indian Army: Its Contribution to the Development of a Nation* Delhi: Oxford University Press.

———. 2004. *The Idea of Pakistan.* Washington, D.C.: Brookings.

Colaresi, Michael. 2001. "Shocks to the System: Great Power Rivalry and the Leadership Long Cycle." *Journal of Conflict Resolution* 45: 569–593.

———. 2004. "When Doves Cry: International Rivalry, Unreciprocated Cooperation, and Leadership Turnover." *American Journal of Political Science* 48, 3: 535–570.

Colaresi, Michael, Karen Rasler, and William R. Thompson. 2007. *Strategic Rivalry: Space, Position, and Conflict Escalation in World Politics.* Cambridge: Cambridge University Press.

Coleman, Peter T. 1997. "Redefining Ripeness: A Social-Psychological Perspective." *Peace and Conflict: Journal of Peace Psychology* 3, 1: 81–103.

Coll, Steve. 2009. "The Back Channel." *New Yorker,* March 2.

Cox, Eric W. 2010. *Why Enduring Rivalries Do—or Don't End.* Boulder, Colo.: Lynne Rienner.

Crittenden, Ann. 1979. "Israel's Economic Plight." *Foreign Affairs* 57, 5 (Summer): 1005–1016.

Cumings, Bruce. 2004. *North Korea.* New York: New Press.

Daigle, Craig. A. 2004. "The Russians Are Going: Sadat, Nixon, and the Soviet Presence in Egypt, 1970–1971." *Middle East Review of International Affairs* 8, 1: 1–15.

DiCicco, Jonathan M. 2004. "War and Rapprochement in the Argentine-Brazilian Rivalry." Paper delivered at the annual meeting of the International Studies Association, March, Montreal.

Diehl, Paul, and Gary Goertz. 2000. *War and Peace in International Rivalry.* Ann Arbor: University of Michigan Press.

Dittmer, Lowell. 1981. "The Strategic Triangle: An Elementary Game Theoretical Analysis." *World Politics* 33: 485–515.

———. 1992. *Sino-Soviet Normalization and Its International Implications, 1945–1990.* Seattle: University of Washington Press.

———. 2004. "The Sino-Russian Strategic Relationship: Ghost of the 'Strategic Triangle'?" In Guoli Li, ed., *Chinese Foreign Policy in Transition.* New York: Aldine de Gruyter.

———. 2011. "The Sino-Russian Partnership: The End of Rivalry?" In Sumit Ganguly and William R. Thompson, eds., *Asian Rivalries: Conflict, Escalation, and Limitations on Two-level Games.* Stanford, Calif.: Stanford University Press.

Dixit, J. N. 2002. *India-Pakistan in War and Peace.* New Delhi: Books Today.

Dogan, Mattei and John Higley. 1998. "Elite, Crises, and Regimes in Comparative Analysis." In Mattei Dogan and John Higley, eds., *Elites, Crises, and the Origins of Regimes.* Lanham, Maryland: Rowman & Littlefield.

Doran, Michael. 1999. *Pan-Arabism before Nasser: Egyptian Power Politics and the Palestine Question.* Oxford: Oxford University Press.

———. 2001. "Egypt." In L. Carl Brown, ed., *Diplomacy in the Middle East: The International Relations of Regional and Outside Powers.* New York: I. B. Tauris Publishers.

Drysdale, Alasdair, and Raymond Hinnebusch. 1991. *Syria and the Middle East Process.* New York: Council on Foreign Relations Press.

Duiker, William J. 1995. *Vietnam: Revolution in Transition.* 2nd ed. Boulder, Colo.: Westview.

Duncan, Peter J. S. 1989. *The Soviet Union and India.* New York: Council on Foreign Relations.

Dutt, Subimal. 1977. *With Nehru in the Foreign Office.* Calcutta: Minerva.

Eban, Abba. 1979. "Camp David: The Unfinished Business." *Foreign Affairs* 57, 2: 343–354.

Eberstadt, Nicholas. 1998. "North Korea's Unification Policy: 1948–1996." In Samuel S. Kim, ed., *North Korean Foreign Relations in the Post–Cold War Era.* Oxford: Oxford University Press.

Elazar, Daniel J. and Shmuel Sandler. 1995. "Introduction: The Battle over Jewishness and Zionism in the Post-Modern Era". In Daniel J. Elazar and Shmuel Sandler, eds., *Israel at the Polls 1996.* London: Frank Cass.

Ellison, Herbert J. 1993. "Soviet-Chinese Relations: The Experience of Two Decades." In Robert S. Ross, ed., *China, the United States, and the Soviet Union: Tripolarity and Policy Making in the Cold War.* Armonk, N.Y.: M. E. Sharpe.

Evangelista, Matthew. 1991. "Sources of Moderation in Soviet Security Policy." In Philip E. Tetlock, Jo L. Husbands, Robert Jervis, Paul C. Stern, and Charles Tilly, eds., *Behavior, Society and Nuclear War.* Vol. 2. New York: Oxford University Press.

Fearon, James. 1998. "Bargaining, Enforcement, and International Cooperation." *International Organization* 52, 2: 269–306.

Feiler, Gil. 2003. *Economic Relations between Egypt and the Gulf Oil States, 1967–2000: Petro-Wealth and Patterns of Influence.* Portland, Ore: Sussex Academic Press.

Fein, Aharon. 1995. "Voting Trends of Recent Immigrants from the Former Soviet Union." In Asher Arian and Michal Shamir, eds., *The Elections in Israel 1992.* Albany: State University of New York Press.

Foot, Rosemary. 1995. *U.S. Relations with China Since 1949.* Oxford: Oxford University Press.

Frazier, Derrick V., and William J. Dixon. 2006. "Third Party Intermediaries and Negotiated Settlements, 1946–2000." *International Interactions* 32, 4: 385–408.

Gaddis, John L. 1992/1993. "International Relations Theory and the End of the Cold War." *International Security* 17, 3: 5–58.

Ganguly, Sumit. 1990. "Deterrence Failure Revisited: The Indo-Pakistani War of 1965." *Journal of Strategic Studies* 13, 4: 77–93.

———. 1994. *The Origins of War in Asia.* 2nd ed. Boulder, Colo.: Westview.

———. 1997. *The Crisis in Kashmir: Portents of War, Hopes of Peace.* New York: Cambridge University Press.

———. 1999. "India's Pathway to Pokhran II: The Sources and Prospects of India's Nuclear Weapons Program." *International Security* 23, 4: 148–177.

———. 2001. *Conflict Unending: India-Pakistani Tension since 1947.* New York: Columbia University Press.

———. 2003. "The Crisis of Indian Secularism." *Journal of Democracy* 14, 4 (October): 11–25.

———. 2005. "The Limits of Coercive Diplomacy in South Asia: The Indo-Pakistani Crisis of 2001–2002." *Security Studies* 14, 2: 290–324.

———. 2006. "Will Kashmir Stop India's Rise?" *Foreign Affairs* 85, 4: 45–56.

———. 2010. "Another India-Pak Impasse." *The Diplomat*, July 20. Available at http://the-diplomat.com/2010/07/20/another-india-pakistan impasse/?print=yes.

Ganguly, Sumit and Devin T. Hagerty. 2005. *Fearful Symmetry: India-Pakistan Crises in the Shadow of Nuclear Weapons.* Seattle: University of Washington Press.

Ganguly, Sumit, and S. Paul Kapur. 2010. *India, Pakistan and the Bomb: Debating Nuclear Stability in South Asia.* New York: Columbia University Press.

Ganguly, Sumit and Michael R. Kraig. 2005. "The 2001–2002 Indo-Pakistani Crisis: Exposing the Limits of Coercive Diplomacy." *Security Studies* 14, 2 (Summer): 290–324.

Ganguly, Sumit, and R. Harrison Wagner. 2004. "India and Pakistan: Bargaining Under the Shadow of Nuclear War." *Journal of Strategic Studies* 27, 3: 479–507.

Garb, Paula, and Susan Allen Nan. 2006. "Negativity in a Correlation Network of Citizen Peacebuilding by Initiatives in the Georgian-Abkahz Peace Process." *International Negotiations* 11, 1: 7–35.

Garver, John W. 1982. *China's Rapprochement with the United States, 1968–1971.* Boulder, Colo.: Westview.

———. 1997. *The Sino-American Alliance: Nationalist China and American Cold War Strategy in Asia*. Armonk, N.Y.: M. E. Sharpe.

Gazit, Shlomo. 1995. *The Carrot and the Stick: Israel's Policy in Judea and Samaria, 1967–68*. Trans. Duevek Danielli. New York: Bnai Brith Books.

Gintis, Herbert, Samuel Bowles, Robert Boyd, and Ernest Fehr, eds. 2005. *Moral Sentiments and Material Interests: The Foundations of Cognition in Economic Life*. Cambridge, Mass.: MIT Press.

Goertz, Gary, and Paul F. Diehl. 1995. "The Initiation and Termination of Enduring Rivalries: The Impact of Political Shocks." *American Journal of Political Science* 39: 30–52.

———. 1997. "The Initiation and Termination of Enduring Rivalaries: The Impact of Political Shocks." *American Journal of Political Science* 39,1: 30–52.

Golan, Galia. 1997. "Moscow and the PLO: The Ups and Downs of a Complex Relationship." In Avraham Sela and Moshe Maoz, eds., *The PLO and Israel: From Armed Revolution to Political Solution, 1964–1994*. New York: St. Martin's Press.

———. 2000. "The Soviet Union and the Yom Kippur War." In P. R. Kumaraswamy, ed., *Revisiting the Yom Kippur War*. London: Frank Cass.

———. 1997. "Moscow and the PLO: The Ups and Downs of a Complex Relationship." In Avraham Sela and Moshe Maoz, eds., *The PLO and Israel: From Armed Revolution to Political Solution, 1964–1994*. New York: St. Martin's Press.

Goldmann, Kjell. 1988. *Change and Stability in Foreign Policy: The Problems and Possibilities of Détente*. Princeton, N.J.: Princeton University Press.

Goldstein, Joshua, and John Freeman. 1990. *Three Way Street: Strategic Reciprocity in World Politics*. Chicago: University of Chicago Press.

Goldstein, Joshua, Jon C. Pevehouse, Deborah J. Gerner, and Shibley Telhami. 1998. "The Dynamics of Middle East Conflict, 1979–90: A Time-Series Analysis." Paper prepared for the 1998 Annual Meeting of the International Studies Association, March, Minneapolis.

———. 2001. "Reciprocity, Triangularity and Cooperation in the Middle East, 1977–97." *Journal of Conflict Resolution* 45, 5: 594–620.

Goldstein, Steven M. 1994. "Nationalism and Internationalism: Sino-Soviet Relations." In Thomas W. Robinson and David Shambaugh, eds., *Chinese Foreign Policy: Theory and Practice*. Oxford: Clarendon Press.

———. 2002. "The Taiwan Strait: A Continuing Status Quo of Deadlock?" *Cambridge Review of International Affairs* 15: 85–94.

Goldstone, Jack. A. 1991. *Revolution and Rebellion in the Early Modern World*. Berkeley: University of California Press.

Gould, Harold A. and Sumit Ganguly. 1992 *The Hope and the Reality: U.S.-Indian Relations from Roosevelt to Reagan*. Boulder, Colo.: Westview.

Grieg, J. Michael, and Paul F. Diehl. 2005. "The Peacekeeping-Peacemaking Dilemma." *International Studies Quarterly* 49, 4: 621–646.

———. 2006. "Softening Up: Making Conflicts More Amenable to Diplomacy." *International Interactions* 32, 4: 315–384.

Groth, Allon. 1995. *The PLO's Road to Peace: Processes of Decision-Making*. London: Royal United Services Institute for Defence Studies.

Gu, Weiqun. 1995. *Conflicts of Divided Nations: The Cases of China and Korea*. Westport, Conn.: Praeger.

Haddad, George M. 1973. *Revolution and Military Rule in the Middle East: The Arab States. Part 2: Egypt, the Sudan, Yemen and Libya*. New York: Robert Speller and Sons.

Haggard, Stephen, and Robert R. Kaufman. 1995. *The Political Economy of Democratic Transitions*. Princeton, N.J.: Princeton University Press.

Halperin, Morton H. 1974. *Bureaucratic Politics and Foreign Policy*. Washington, D.C.: Brookings Institution.

Harding, Harry. 1992. *A Fragile Relationship: the United States and China Since 1972*. Washington, D.C.: Brookings Institution.

Hassassian, Manuel S. 1997. "Policy and Attitude Changes in the Palestine Liberation Organization, 1965–1994." In Avraham Sela and Moshe Maoz, eds., *The PLO and Israel: From Armed Revolution to Political Solution, 1964–1994*. New York: St. Martin's Press.

Hartzell, Caroline A. 1999. "Explaining the Stability of Negotiated Settlements to Intrastate Wars." *Journal of Conflict Resolution* 43, 1: 3–22.

Heo, Uk, and Terence Roehrig. 2010. *South Korea since 1980*. Cambridge: Cambridge University Press.

Herz, John H. 1950. "Idealist Internationalism and the Security Dilemma." *World Politics* 2, 2:157–180.

Hermann, Charles F. 1990. Changing Course: When Governments Choose to Redirect Foreign Policy." *International Studies Quarterly* 34, 1: 3–21.

Herrmann, Richard, and Richard Ned Lebow, eds. 2004. *Ending the Cold War: Interpretations, Causations, and the Study of International Relations*. New York: Palgrave-Macmillan.

Hindustan Times. 2010. "Pillai Remark Was Ill-Timed, Says Krishna." *Hindustan Times*, July 21.

Hinnebusch, Raymond A. 1996. "Does Syria Want Peace? Syrian Policy in the Syrian-Israeli Peace Negotiations." *Journal of Palestine Studies* 26, 1: 42–57.

Hoare, James E., and Susan Pares. 1999. *Conflict in Korea: An Encyclopedia*. Santa Barbara, Calif.: ABC-CLIO.

Hoffman, Steven. 1972. "Anticipation, Disaster and Victory India: 1962–71." *Asian Survey* 12, 11: 960–979.

———. 1990. *India and the China Crisis*. Berkeley: University of California.

Holsti, K. J., Thomas W. Robinson, Timothy M. Shaw, Miguel Monterichard, and Ibrahim Msabaha. 1982. *Why Nations Realign: Foreign Policy Restructuring in the Postwar World*. Winchester, Mass.: Allen and Unwin.

Holsti, Ole R. 1967. "Cognitive Dynamics and Images of the Enemy." In John C. Farrell and Asa P. Smithy, eds., *Image and Reality in World Politics*. New York: Columbia University.

Hood, Steven J. 1992. *Dragon Entangled: Indochina and the China-Vietnam War*. Armonk, N.Y.: M. E. Sharpe.

Hopwood, Derek. 1991. *Egypt: Politics and Society*. London: Routledge.

Horn, Robert C. 1982. *Soviet Indian Relations: Issues and Influence*. New York: Praeger.

Huang, Chi, Woosang Kim, and Samuel S. G. Wu. 1995. "Rivalry between the ROC and PRC: An Expected Utility Theoretical Perspective." In Tun-jen Cheng, Chi Huang, and Samuel S. G. Wu, eds., *Inherited Rivalry: Conflict across the Taiwan Straits*. Boulder, Colo.: Lynne Rienner.

Hunter, F. Robert. 1993. *The Palestinian Uprising: A Way by Other Means*. 2nd ed. Berkeley: University of California Press.

Husain, Javid. 2010. "Deadlock in India-Pakistan Talks." *The Nation* (Islamabad), March 2.

Hutchinson, Sharon E. 2001. "A Curse from God? Religious and Political Dimensions of the Post-1991 Rise of Ethnic Violence in South Sudan." *Journal of Modern African Studies* 39, 2: 207–331.

Huth, Paul K. 1996. *Standing Your Ground: Territorial Disputes and International Conflict*. Ann Arbor: University of Michigan Press.

Ilahi, Shereen Ilahi. 2003. "The Radcliffe Boundary Commission and the Fate of Kashmir." *India Review* 2, 1: 77–102.

Inbar, Efraim. 1991. *War and Peace in Israeli Politics: Labor Party Positions on National Security*. Boulder, Colo.: Lynn Rienner.

———. 1999. *Rabin and Israel's National Security*. Baltimore: Johns Hopkins University Press.

Indian Express. 2010. "The ISI Controlled 26/11 Attacks from Beginning to End." *Indian Express*, July 17. Available at http://www.indianexpress.com/news/the-isi-controlled-coordinated-26-11-attacks-from-beginning-to-end/647849/0.

Jackson, Robert. 1975. *South Asian Crisis: India, Pakistan, Bangla Desh*. London: Chatto and Windus.

Jacob, J. F. R. 1997. *Surrender at Dhaka: The Birth of a Nation*. New Delhi: Manohar Publishers.

Jayanth, Jacob. 2010. "July 16: Cruel Date for Indo-Pak Ties." *Hindustan Times*, July 17.

Jervis, Robert. 1976. *Perception and Misperception in International Politics*. Princeton, N.J.: Princeton University Press.

———. 1997. *System Effects: Complexity in Political and Social Life*. Princeton, N.J.: Princeton University Press.

Johnson, Rob. 2005. *A Region in Turmoil: South Asian Conflicts since 1947*. London: Reaktion Books.

Jones, Owen Bennett. 2002. *Pakistan: Eye of the Storm.* New Haven, Conn.: Yale University Press.

Kahn, Joseph. 2004. "China's Saber Rattling: Paying Off?" *New York Times*, December 13, A8.

Kang, C. S. 2003a. "North Korea's International Relations: The Successful Failure?" In Samuel S. Kim, ed., *The International Relations of Northeast Asia.* Lanham, Md.: Rowman and Littlefield.

———. 2003b. "Acute Conflicts in Asia after the Cold War: Kashmir, Taiwan and Kashmir." In Muthiah Alagappa, ed., *Asian Security Order: Instrumental and Normative Features.* Stanford, Calif.: Stanford University Press.

Karawan, Ibrahim. 2005. "Foreign Policy Restructuring: Egypt's Disengagement from the Arab-Israeli Conflict Revisited." *Cambridge Review of International Affairs* 18, 3: 326–338.

Kastner, Scott L. 2007. *Political Conflict and Economic Interdependence across the Taiwan Strait and Beyond.* Stanford, Calif.: Stanford University Press.

Kelman, Herbert C. 1995. "Contributions of an Unofficial Conflict Resolution Effort to the Israeli-Palestinian Breakthrough." *Negotiation Journal* (January): 19–27.

Kenawy, Ezzat Molouk. 2009. "The Economic Development in Egypt during the 1952–2007 Period." *Australian Journal of Basic and Applied Sciences* 3, 2: 588–603.

Kessler Neff, Martha. 2000. "Syria, Israel and the Middle East Process: Past Success and Final Challenge." *Middle East Policy* 7, 2: 68–89.

Khan, Akbar. 1970. *Raiders in Kashmir.* Karachi: Pak Publishers.

Kim, Abraham. 2004. "The Challenges of Peacefully Reunifying the Korean Peninsula." In Samuel S. Kim, ed., *Inter-Korean Relations: Problems and Prospects.* New York: Palgrave-Macmillan.

Kim, Samuel S. 1994. *Inter-Korean Relations: Problems and Prospects.* New York: Palgrave Macmillan.

———. 2011. "The Rivalry Between the Two Koreas." In Sumit Ganguly and William R. Thompson, eds., *Asian Rivalries: Conflict, Escalation, and Limitations on Two-Level Games.* Stanford, Calif.: Stanford University Press.

Kingdon, John W. 1995. *Agendas, Alternatives and Public Policies.* 2nd ed. New York: Harper Collins.

Knight, Alan. 1998. "Historical and Theoretical Considerations." In Mattei Dogan and John Higley, eds., *Elites, Crises, and the Origins of Regimes.* Lanham, Md.: Rowman & Littlefield.

Korn, David A. 1992. *Stalemate: The War of Attrition and Great Power Diplomacy in the Middle East, 1967–1970.* Boulder, Colo.: Westview.

Krasner, Stephen D. 1984. "Approaches to the State: Alternative Conceptions and Historical Dynamics." *Comparative Politics* 16: 223–246.

Kriesberg, Louis. 1991. *Timing the De-Escalation of International Conflicts.* Syracuse, N.Y.: Syracuse University Press.

———. 992. *International Conflict Resolution: TheU.S.-U.S.S.R. and Middle East Cases.* New Haven, Conn.: Yale University Press.

———. 1998. *Constructive Conflicts: From Escalation to Resolution.* Lanham, Md.: Rowman and Littlefield.

———. 2007. *Constructive Conflicts: From Escalation to Resolution.* 3rd ed. Lanham, Md.: Rowman and Littlefield.

Krishna, Ashok, and P. R. Chari, eds. 2001. *Kargil: The Tables Turned.* New Delhi: Manohar.

Kumar, Radha. 2005. *Making Peace with Partition.* New Delhi: Penguin Books.

Kupchan, Charles A. 2010. *How Enemies Become Friends: The Sources of Stable Peace.* Princeton, N.J.: Princeton University Press.

Kydd, Andrew H. 2005. *Trust and Mistrust in International Relations.* Princeton, N.J.: Princeton University Press.

Lamb, Alastair. 1991 . *Kashmir: A Disputed Legacy, 1846–1990.* Karachi: Oxford University Press.

Laron, Guy. 2009. "'Logic Dictates That They May Attack When They Feel They Can Win': The 1955 Czech-Egyptian Arms Deal, the Egyptian Army, and Israeli Intelligence." *Middle East Journal* 63, 1 (Winter): 69–84.

Lavoy, Peter. 2009. *Asymmetric Warfare in South Asia: The Causes and Consequences of the Kargil Conflict.* Cambridge: Cambridge University Press.

Lebow, Richard Ned. 1981. *Between Peace and War: The Nature of International Crisis.* Baltimore, Md.: Johns Hopkins University Press.

———. 1995. "The Search for Accommodation: Gorbachev in Comparative Perspective." In Richard Ned Lebow and Thomas Risse-Kappen, eds., *International Relations Theory and the End of the Cold War.* New York: Columbia University Press.

———. 1997. "Transitions and Transformations: Building International Cooperation." *Security Studies* 6: 154–79.

Lebow, Richard Ned, and Thomas Risse-Kappen, eds. 1995. *International Relations Theory and the End of the Cold War.* New York: Columbia University Press.

Legro, Jeffrey W. 2005. *Rethinking the World: Great Power Strategies and International Order.* Ithaca, N.Y.: Cornell University Press.

Legvold, Robert. 1993. "Sino-Soviet Relations: the U.S. Factor." In Robert S. Ross, ed., *China, the United States, and the Soviet Union: Tripolarity and Policy Making in the Cold War.* New York: M. E. Sharpe.

Leifer, Michael. 1995. "Vietnam's Foreign Policy in the Post-Soviet Era: Coping with Vulnerability." In Robert S. Ross, ed., *East Asia in Transition: Toward a New Regional Order.* Armonk, N.Y.: M. E. Sharpe.

Leng, Russell J. 1983. "When Will They Ever Learn? Coercive Bargaining in Recurrent Crises." *Journal of Conflict Resolution* 27: 379–419.

Levy, Jack S. 1994 ."Learning and Foreign Policy: Sweeping a Conceptual Minefield." *International Organization* 48: 279–312.

Levy, Yagil. 1997. *Trial and Error: Israel's Route from War to De-Escalation*. Albany: State University of New York Press.

Li Jie. 2001. "Changes in China's Domestic Situation in the 1960s and Sino-U.S. Relations." In Robert S. Ross and Changbin Jiang, eds., *Re-examining the Cold War: U.S.-China Diplomacy, 1954–1973*. Cambridge, Mass.: Harvard University Asia Center.

Lichbach, Mark I. 1995. *The Rebel's Dilemma*. Ann Arbor: University of Michigan Press.

Licklider, Roy, 1993. "What We Have Learned and Where Do We Go from Here?" In Roy Licklider, ed., *Stopping the Killing: How Civil Wars End*. New York: New York University Press.

Long, William J., and Peter Brecke. 2003. *War and Reconciliation: Reason and Emotion in Conflict Resolution*. Cambridge, Mass.: MIT Press.

Magnier, Mark. 2010. "First Official Meeting between Pakistan, India since Mumbai Attack Is Strained." *Los Angeles Times*, February 25.

Makovsky, David. 1996. *Making Peace with the PLO: The Rabin Government's Road to the Oslo Accord*. Boulder, Colo.: Westview Press.

Mani, Kristina. 2011. *Democratization and Military Transformation in Argentina and Chile: Rethinking Rivalry*. Boulder, Colo.: Lynne Rienner.

Mann, Jim. 1999. *About Face: A History of America's Curious Relationship with China from Nixon to Clinton*. New York: Alfred Knopf.

Ma'oz, Moshe. 1995. *Syria and Israel: From War to Peacemaking*. Oxford: Clarendon Press.

———. 2002. "Syrian-Israeli Relations on the Eve of the Peace Process." In Joseph Ginant, Edward J. Perkins, and Edwin G. Corr, eds., *The Middle East Process: Vision versus Reality*. Norman: University of Oklahoma Press.

Maoz, Zeev. 2011. *Networks of Nations: The Evolution, Structure, and Impact of International Networks, 1816–2001*. Cambridge: Cambridge University Press.

Maoz, Zeev, and Ben D. Mor. 2002. *Bound by Struggle: Strategic Evolution of Enduring International Rivalries*. Ann Arbor: University of Michigan Press.

Mauryin, Mark E. 2002. "North-South Korean Relations: A Chronology of Events in 2000 and 2001." Washington, D.C.: Congressional Research Service, Library of Congress.

Maxwell, Neville. 1970. *India's China War*. London: Cape.

Mayer, Thomas. 1983. *Egypt and the Palestine Question 1936 to 1945*. Berlin: Klaus Schwarz Verlag.

McAdam, Doug, Sidney Tarrow, and Charles Tilly. 1997. "Toward an Integrated Perspective on Social Movements and Revolutions." In Mark I. Lichbach and Alan S. Zuckerman, eds., *Comparative Politics: Rationality, Culture and Structure*. Cambridge: Cambridge University Press.

McGrath, Allen. 1996. *The Destruction of Pakistan's Democracy*. New York: Oxford University Press.

McMahon, Robert J. 1994. *The Cold War on the Periphery: The United States, India, and Pakistan*. New York: Columbia University Press.

McNamara, Robert. 2003. *Britain, Nasser and the Balance of Power in the Middle East 1952–1967*. London: Frank Cass.

Mendelson, Sarah E., 1998. *Changing Course: Ideas, Politics, and the Soviet Withdrawal from Afghanistan*. Princeton, N.J.: Princeton University Press.

Mitchell, Christopher R. 1995. "Asymmetry and Strategies of Regional Conflict Reduction." In William Zartman and Victor A. Kremenyuk, eds., *Cooperative Security: Reducing Third World Wars*. New York: Syracuse University Press.

Modelski, George. 1996. "Evolutionary Paradigm for Global Politics." *International Studies Quarterly* 40: 321–42.

Mohan, C. Raja. 2004. *Crossing the Rubicon: The Shaping of India's New Foreign Policy* New York: Palgrave Macmillan.

Moon, Chung-in. 1997. "The North Korean Problem and the Role of South Korea." In Thomas H. Henricksen and Jongryn Mo, eds., *North Korea after Kim Il Sung: Continuity or Change?* Stanford, Calif.: Hoover Institution Press.

Moon, Chung-in, and Taehwan Kim. 2003. "South Korea's International Relations: Challenges to Developmental Realism?" In Samuel S. Kim, ed., *The International Relations of Northeast Asia*. Lanham, Md.: Rowman and Littlefield.

Morris, Benny. 1993. *Israel's Border Wars 1949–1956*. Oxford: Clarendon Press.

Muslih, Muhammed. 1997. "A Study of PLO Peace Initiatives, 1974–1988." In Avraham Sela and Moshe Maoz, eds., *The PLO and Israel: From Armed Revolution to Political Solution, 1964–1994*. New York: St. Martin's Press.

Myers, Ramon H., and Jialin Zhang. 2006. *The Struggle across the Taiwan Strait: The Divided China Problem*. Palo Alto, Calif.: Hoover Institution Press.

Nanto, Dick K. 2003. "North Korea: Chronology of Provocations, 1950–2003." Report for Congress, Congressional Research Services, Library of Congress.

Neff, Donald. 1981. *Warriors at Suez: Eisenhower takes America into the Middle East*. New York: Linden Press/Simon and Schuster.

———. 1994. "Israel-Syria: Conflict at the Jordan River, 1949–1967." *Journal of Palestine Studies* 23, 4 (Summer): 26–40.

Nelson, Harvey W. 1989. *Power and Insecurity: Beijing, Moscow, and Washington, 1949–1988*. Boulder, Colo.: Lynne Rienner.

Ninh, Kim. 1998. "Vietnam: Struggle and Cooperation." In Muthiah Alagappa, ed., *Asian Security Practice: Material and Ideational Influences*. Stanford, Calif.: Stanford University Press.

Nisbet, Robert A. 1969. *Social Change and History: Aspects of the Western Theory of Development*. New York: Oxford University Press.

Noland, Marcus. 1997. "Prospects for a North Korean External Economic Opening." In Thomas H. Henricksen and Jongryn Mo, eds., *North Korea after Kim Il Sung: Continuity or Change?* Stanford, Calif.: Hoover Institution Press.

Noorani, A. G. 2005. "The Truth about Agra." *Frontline* 22 (July 16–29): 15.

Oberdorfer, Don. 2001. *The Two Koreas: A Contemporary History*. Rev. and updated ed. New York: Basic Books.

Oberschall, Anthony. 1996. "Opportunities and Framing in the Eastern European Revolts of 1989." In Doug McAdam, John D. McCarthy, and Mayer N. Zald, eds., *Comparative Perspectives on Social Movements: Political Opportunities, Mobilizing Structures and Cultural Framing*. Cambridge: Cambridge University Press.

Oren, Michael. 1992. *Origins of the Second Arab-Israel War: Egypt, Israel and the Great Powers 1952–56*. London: Frank Cass.

Orme, John D. 2004. *The Paradox of Peace: Leaders, Decisions, and Conflict Resolution*. New York: Palgrave Macmillan.

Palit, D. K. 1992. *War in the High Himalaya: The Indian Army in Crisis*. London: C. Hurst and Company.

Paul, T. V., ed., 2005. *The Indo-Pakistani Conflict: An Enduring Rivalry*. New York: Cambridge University Press.

Pervin, David J. 1997. "Building Order in Arab-Israeli Relations: From Balance to Concert?" In David A. Lake and Patrick M. Morgan, eds., *Regional Orders: Building Security in a New World*. University Park: Pennsylvania State University Press.

Podeh, Elie. 2004. "Demonizing the Other: Israel's Perceptions of Nasser and Nasserism." In Elie Podeh and Onn Winckler, eds., *Rethinking Nasserism: Revolution and Historical Memory in Modern Egypt*. Gainesville: University of Florida Press.

Porter, Gareth. 1993. *Vietnam: The Politics of Bureaucratic Socialism*. Ithaca, N.Y.: Cornell University Press.

Pruitt, Dean C. 1981. "Kissinger as a Traditional Mediator with Power." In Jeffrey Z. Rubin,ed., *Dynamics of Third Party Intervention: Kissinger in the Middle East*. New York: Praeger.

Rabasa, Angel, Robert D. Blackweill, Peter Chalk, Kim Cragin, C. Christine Fair, Brian A. Jackson, Brian Michael Jenkins, Seth G. Jones, Nathaniel Shestak, and Ashley J. Tellis. 2009. *The Lessons of Mumbai*. Santa Monica, Calif.: Rand Corporation.

Rabinovich, Itamar. 1991. *The Road Not Taken: Early Arab-Israeli Negotiations*. Oxford: Oxford University Press.

———. 1998. *The Brink of Peace: The Israeli-Syrian Negotiations*. Princeton, N.J.: Princeton University Press.

———. 2004. *Waging Peace: Israel and the Arabs 1948–2003*. Princeton, N.J.: Princeton University Press.

Rajmaira, Sheen. 1997. "Indo-Pakistani Relations: Reciprocity in Long-term Perspective." *International Studies Quarterly* 41, 3: 547–560.

Rasler, Karen. 2001a. "Political Shocks and the Deescalation of Protracted Conflicts: The Israeli-Palestinian Case." In William R. Thompson, ed., *Evolutionary Interpretations of World Politics*. New York: Routledge.

———. 2001b. "Shocks, Expectancy Revision and the De-escalation of Protracted Conflicts: The Israeli-Palestinian Case." *Journal of Peace Research* 37: 699–720.

Rasler, Karen and William R. Thompson. 1989. *War and State Making: The Shaping of the Global Powers.* Boston: Unwin Hyman.

Regan, Patrick M. 2002. "Third-Party Interventions and the Duration of Intrastate Conflict." *Journal of Conflict Resolution* 46, 1: 55–73.

Regan, Patrick M. and Allan C. Stam. 2000. "In the Nick of Time: Conflict Management, Mediation Timing, and the Duration of Interstate Disputes." *International Studies Quarterly* 44, 2: 239–260.

Reiter, Dan. 1994. "Learning, Realism, and Alliances: The Weight of the Shadow of the Past." *World Politics* 46, 2: 490–526.

Rivlin, Paul. 1997. "Leadership and the Economy in the Arab World, 1950–1996." *Middle East Review of International Affairs.* 1, 3 (September).

Rivlin, Paul, and Schmuel Even. 2004. *Political Stability in Arab States: Economic Causes and Consequences.* Memorandum 74 (December). Tel Aviv: Jaffee Center for Strategic Studies, Tel Aviv University

Robinson, Glenn E. 1997. *Building a Palestinian State: The Incomplete Revolution.* Bloomington: Indiana University Press.

Robinson, Thomas W. 1994. "Chinese Foreign Policy from the 1940s to the 1990s." In Thomas W. Robinson and David Shambaugh, eds., *Chinese Foreign Policy: Theory and Practice.* Oxford: Clarendon Press.

Rock, Stephen R. 1989. *Why Peace Breaks Out: Great Power Rapprochement in Historical Perspective.* Chapel Hill: University of North Carolina Press.

Rosati, Jerel A. 1994. "Cycles in Foreign Policy Restructuring: The Politics of Continuity and Change in U.S. Foreign Policy." In Jerel A. Rosati, Joe D. Hagan, and Martin W. Sampson III, eds., *Foreign Policy Restructuring: How Governments Respond to Global Change.* Columbia: University of South Carolina Press.

Rosenau, James N. 1981. *The Study of Political Adaptation: Essays on the Analysis of World Politics.* New York: Nichols.

Ross, Robert S. 1988. *The Indochina Tangle: China's Vietnam Policy, 1975–1979.* New York: Columbia University Press.

———. 1995. *Negotiating Cooperation: The United States and China, 1969–1989.* Stanford, Calif.: Stanford University Press.

———. 2001. "Beijing as a Conservative Power." *Foreign Affairs* 76, 2: 33–44.

Rozman, Gilbert. 2010. *Chinese Strategic Thought toward Asia.* New York: Palgrave-Macmillan.

Rubin, Barry. 1994. "Iraq and the PLO: Brother's Keepers, Losers Weepers." In Amatzia Baram and Barry Rubin, eds, *Iraq's Road to War.* London: Macmillan.

Sachar, Howard. 1981. *Egypt and Israel.* New York: Richard Marek Publishers.

Shachar, Ron and Michal Shamir. 1995. "Modelling Victory in the 1992 Election." In Asher Arian and Michal Shamir, eds., *The Elections in Israel 1992.* Albany: New York: State University of New York Press.

Sahliyeh, Emile F. 1986. *The PLO after the Lebanon War.* Boulder, Colo.: Westview Press.

Sar Desai, D. R. 1994. *Southeast Asia: Past and Present*. Boulder, Colo.: Westview.

Sattar, Abdul. 1950. "Simla Pact: Negotiation Under Duress." *Regional Studies* (Islamabad) 13, 3 (Autumn).

Sayigh, Yezid. 1997. *Armed Struggle and the Search for a State: The Palestinians National Movement, 1949–1993*. Oxford: Clarendon Press.

Schaller, Michael. 2002. *The United States and China: Into the 21st Century*. Oxford: Oxford University Press.

Schmidt Adams, Dana. 1968. *Yemen: The Unknown War*. New York: Holt, Rinehart, and Winston.

Schrodt, Philip and Deborah A. Gerner. 2004. "An Event Data Analysis of Third Party Mediation." *Journal of Conflict Resolution* 48, 3: 31–330.

Schulzinger, Robert D. 2001. "The Johnson Administration, China and the Vietnam War." In Robert S. Ross and Changbin Jiang, eds., *Re-Examining the Cold War: U.S.-China Diplomacy, 1954–1973*. Cambridge, Mass.: Harvard University Asia Center.

Scobell, Andrew. 2004. "China and Inter-Korean Relations: Beijing as Balancer." In Samuel S. Kim, ed., *Inter-Korean Relations: Problems and Prospects*. New York: Palgrave-Macmillan.

Scott, David. 2007. *China Stands Up: The PRC and the International System*. London: Routledge.

Seale, Patrick. 1988. *Asad: The Struggle for the Middle East*. Berkeley: University of California Press.

———. 2000. "The Syria-Israel Negotiations: Who Is Telling the Truth?" *Journal of Palestine* 29, 2: 65–77.

Seale, Patrick, and Linda Butler. 1996. "Asad's Regional Strategy and the Challenge from Netanyahu." *Journal of Palestine Studies* 26, 1: 27–41.

Segev, Samuel. 1998. *Crossing the Jordan: Israel's Hard Road to Peace*. New York: St. Martin's Press.

Sela, Avraham. 1998. *The Decline of the Arab-Israeli Conflict: Middle East Politics and the Quest for Regional Order*. Albany: State University of New York Press.

———. 2000. "The 1973 Arab War Coalition: Aims, Coherence, and Gain-Distribution." In P. R. Kumaraswamy, ed., *Revisiting the Yom Kippur War*. London: Frank Cass.

Shambaugh, David. 1994. "Patterns of Interaction in Sino-American Relations." In Thomas W. Robinson and David Shambaugh, eds., *Chinese Foreign Policy: Theory and Practice*. Oxford: Clarendon Press.

Sharma, Ashok. 2010. "Pakistan and India Resume Talks Halted after Mumbai Attacks." *The Guardian*, April 29.

Sharp, Jeremy M. 2009. "U.S. Foreign Aid to Israel." Congressional Research Service, CRS Report for Congress. www.crs.gov.

Shemesh, Moshe. 2008. "The Origins of Sadat's Strategic Volte-Face." *Israel Studies* 13, 2: 28–53.

Sheng, Lijun. 2002. "The Taiwan Issue: Does China Have a Strategy?" *Cambridge Review of International Affairs* 15: 123–30.

Shih, Chih-yu. 1995. "The Reluctant Dragon: Taiwan as a Psychological Factor in Mainland China's U.S. Policy." In Tun-jen Cheng, Chi Huang, and Samuel S. G. Wu, eds., *Inherited Rivalry: Conflict across the Taiwan Straits*. Boulder, Colo.: Lynne Rienner.

Shimshoni, Jonathan. 1988. *Israel and Conventional Deterrence: Border Warfare from 1953 to 1970*. Cornell Studies in Security Affairs. Ithaca, N.Y.: Cornell University Press.

Shlaim, Avi. 2001a. *The Iron Wall: Israel and the Arab World*. New York: Norton and Company.

———. 2001b. "Israel and the Arab Coalition in 1948." In Eugene L. Rogan and Avi Shlaim, eds., *The War for Palestine: Rewriting the History of 1948*. Cambridge: Cambridge University Press.

Singh, Amarinder. 2001. *A Ridge Too Far: War in the Kargil Heights, 1999*. Patiala: Motibagh Palace.

Singh, Surinder Nihal. 1986. *The Yogi and the Bear: Story of Indo-Soviet Relations*. New Delhi: Allied Publishers.

Sisson, Richard, and Leo E. Rose. 1991. *War and Secession: Pakistan, India and the Creation of Bangladesh*. Berkeley: University of California Press.

Skocpol, Theda. 1979. *States and Social Revolutions: A Comparative Analysis of France, Russia and China*. Cambridge: Cambridge University Press.

Slater, Jerome. 2002. "Lost Opportunities for Peace in the Arab Israeli Conflict: Israel and Syria, 1948–2001." *International Security* 27, 1: 79–106.

Snyder, Scott. 2004. "Inter-Korean Relations: A South Korean Perspective." In Samuel S. Kim, ed., *Inter-Korean Relations: Problems and Prospects*. New York: Palgrave-Macmillan.

———. 2010. "Changes in Seoul's North Korean Policy and Implications for Pyongyang's Inter-Korean Diplomacy." In Kyung Ae Park, ed., *New Challenges of North Korean Foreign Policy*. New York: Palgrave-Macmillan.

Stein, Janice. 1985. "Structures, Strategies, and Tactics of Mediation: Kissinger and Carter in the Middle East." *Negotiation Journal* 1, 4: 331–347.

———. 1993a. "The Political Economy of Security Agreements: The Linked Costs of Failure at Camp David." In Peter B. Evans, Harold K. Jacobson, and Robert D. Putnam, eds., *International Bargaining and Domestic Politics: Double-Edge Diplomacy*. Berkeley: University of California Press.

———. 1993b. "International Co-operation and Loss Avoidance: Framing the Problem." In Janice Gross Stein and Louis W. Pauly, eds., *Choosing to Co-operate: How States Avoid Loss*. Baltimore: Johns Hopkins University Press.

Stein, Kenneth W. 1999. *Heroic Diplomacy: Sadat, Kissinger, Carter, Begin and the Quest for Arab-Israeli Peace*. New York: Routledge.

Stone, David. 2004. *Wars of the Cold War: Campaigns and Conflicts, 1945–1990*. London: Brassey's UK .

Stuart-Fox, Martin. 2003. *A Short History of China and Southeast Asia: Tribute, Trade and Influence*. Crow's Nest, NSW (Australia): Allen and Unwin.

Su, Chi. 1993. *Taiwan's Relations with Mainland China: A Tail Wagging Two Dogs*. London: Routledge.

Sutter, Karen. 1993. "China's Vietnam Policy: The Road to Normalization and Prospects for the Sino-Vietnamese Relationship." *Journal of Northeast Asian Studies* 12: 21–46.

Sutter, Robert G. 2008. *Chinese Foreign Relations: Power and Policy since the Cold War*. Lanham, Md.: Rowman and Littlefield.

Swami, Praveen. 1999. *The Kargil War*. New Delhi: Leftword Books.

Tal, David.1996. "Israel's Road to the 1956 War." *International Journal of Middle East Studies* 28, 1 (February): 59–81.

Talbott, Strobe. 2006. *Engaging India: Diplomacy, Democracy and the Bomb*. Washington, D.C.: Brookings Institution Press.

Telhami, Shibley. 1996. "Israeli Foreign Policy: A Realist Ideal-Type or a Breed of Its Own?" In Michael N. Barnett, ed., *Israel in Comparative Perspective: Challenging the Conventional Wisdom*. New York: State University of New York Press.

Thomas, Raju G. C. 1986. *Indian Security Policy*. Princeton, N.J.: Princeton University Press.

Thompson, William R. 2001a. "Identifying Rivals and Rivalries in World Politics." *International Studies Quarterly* 45: 557–86.

———. 2001b. "Expectancy Theory, Strategic Rivalry Deescalation, and the Evolution of the Sino-Soviet Case." In William R. Thompson, ed., *Evolutionary Interpretations of World Politics*. New York: Routledge.

———. 2003. "A Streetcar Named Sarajevo: Catalysts, Multiple Causation, Chains and Rivalry Strategies." *International Studies Quarterly* 47, 3: 453–474.

———. 2005. "Explaining Rivalry Termination in Contemporary Eastern Eurasia with Evolutionary Expectancy Theory." Paper delivered at the International Studies Association, Honolulu, March 3–7.

Thompson, William R. and David R. Dreyer. 2011. *Handbook of International Rivalries, 1494–2010*. Washington, D.C.: Congressional Quarterly Press.

Tilly, Charles, 1978. *From Mobilization to Revolution*. Reading, Mass.: Addison-Wesley.

Tisdell, Simon. 2010. "India and Pakistan's Proxy War Puts Afghanistan Exit at Risk." *The Guardian*, May 7.

Touval, Saadia. 1975. "Biased Intermediaries: Theoretical and Historical Considerations." *Jerusalem Journal of International Relations* 1: 51–69.

———. 1982. *The Peace Brokers*. Princeton, N.J.: Princeton University Press.

Touval, Saadia and I. William Zartman. 1985. "Introduction: Mediation in Theory." In Saadia Touval and I. William Zartman, eds., *International Mediation in Theory and Practice*. Boulder, Colo.: Westview Press.

Tuker, Francis. 1950. *While Memory Serves*. London: Cassell.

Tyler, Patrick. 1999. *A Great Wall: Six Presidents and China: An Investigative History.* New York: Public Affairs.

Unger, Danny. 1995. "From Domino to Dominant: Thailand's Security Policies in the Twenty-First Century." In Robert S. Ross, ed., *East Asia in Transition: Toward a New Regional Order.* Armonk, N.Y.: M. E. Sharpe.

U.S. Library of Congress. 1988. "Israel: A Country Study." Federal Research Division, Library of Congress. Ed. Helen Chapin Metz. Washington, D.C.: Federal Research Division. http://purl.access.gpo.gov.ezproxy.lib.indiana.edu/GPO/LPS40309. Accessed July 10, 2012.

Volgy, Thomas J. and John E. Schwarz. 1994. "Foreign Policy Restructuring and the Myriad Webs of Restraint." In Jerel A. Rosati, Joe D. Hagan, and Martin W. Sampson III, eds., *Foreign Policy Restructuring: How Governments Respond to Global Change*. Columbia: University of South Carolina Press.

Wachman, Alan. 2007. *Why Taiwan? Geostrategic Rationales for China's Territorial Integrity.* Stanford, Calif.: Stanford University Press.

Walter, Barbara F. 1999. "Designing Transitions from Civil War." In Barbara F. Walter and Jack Snyder, eds., *Civil Wars, Insecurity, and Intervention*. New York: Columbia University Press.

Wang, Jianwei. 2003. "Territorial Disputes and Asian Security: Sources, Management, and Prospects." In Muthiah Alagappa, ed., *Asian Security Order: Instrumental and Normative Features*. Stanford, Calif.: Stanford University Press.

Ward, Michael D. 1982. "Cooperation and Conflict in Foreign Policy Behavior." *International Studies Quarterly* 26: 87–126.

Waterbury, John. 1983. *The Egypt of Nasser and Sadat: The Political Economy of Two Regimes.* Princeton, N.J.: Princeton University Press.

Waterman, Harvey. 1993. "Political Order and the Settlement of Civil Wars." In Roy Licklider, ed., *Stop the Killing: How Civil Wars End.* New York: New York University Press.

Wattanayagorn, Panitan. 1998. "Thailand: The Elite's Shifting Conceptions of Security." In Muthiah Alagappa, ed., *Asian Security Practice: Material and Ideational Influences*. Stanford, Calif.: Stanford University Press.

Weible, Christopher M., and Paul A. Sabatier. 2006. "A Guide to the Advocacy Coalition Framework." In Gerald Miller and Mara S. Sidney, eds., *Handbook of Public Policy Analysis: Theory, Politics, and Methods*. Boca Raton, Fla.: CRC Press.

Welch, David A. 2005. *Painful Choices: A Theory of Policy Change*. Princeton, N.J.: Princeton University Press.

Westad, Odd Arne. 2006. *The Third Indochina War: Conflict between China, Vietnam, and Cambodia, 1972–79.* London: Routledge.

Wishnick, Elizabeth. 2004. "Russia in Inter-Korean Relations." In Samuel S. Kim, ed., *Inter-Korean Relations: Problems and Prospects*. New York: Palgrave-Macmillan.

Womack, Brantly. 2006. *China and Vietnam: The Politics of Asymmetry.* Cambridge: Cambridge University Press.

——. 2011. "Asymmetric Rivals: China and Vietnam." In Sumit Ganguly and William R. Thompson, eds., *Asian Rivalries: Conflict, Escalation and Limitation on Two-Level Games*. Stanford, Calif.: Stanford University Press.

Wyatt, David K. 2003. *A Short History of Thailand*. 2nd ed. New Haven, Conn.: Yale University Press.

Yaqub, Salim. 2007. "The Politics of Stalemate: The Nixon Administration and the Arab-Israeli Conflict, 1969–73." In Nigel J. Ashton, ed., *The Cold War in the Middle East: Regional Conflict and the Superpowers, 1967–73*. New York: Routledge.

Young, H. Peyton. 1998. *Individual Strategy and Social Structure: An Evolutionary Theory of Institutions*. Princeton, N.J.: Princeton University Press.

Zacek, Jane Shapiro. 1998. "Russia in North Korean Foreign Policy." In Samuel S. Kim, ed., *North Korean Foreign Relations in the Post–Cold War Era*. Oxford: Oxford University Press.

Zaheer, Hasan. 1994. *The Separation of East Pakistan: The Rise and Realization of Bengali Muslim Nationalism*. Karachi: Oxford University Press.

Zartman, I. William. 1993. "The Unfinished Agenda: Negotiating Internal Conflicts." In Roy Licklider, ed., *Stopping the Killing: How Civil Wars End*. New York: New York University Press.

——. 2001. "The Timing of Peace Initiatives: Hurting Stalemates and Ripe Moments." *Global Review of Ethnopolitics* 1, 1: 8–18.

Zartman, I. William and Johannes Aurik. 1991. "Power Strategies in De-Escalation." In Louis Kriesberg and Stuart Thorson, eds., *Timing the De- Escalation of International Conflicts*. New York: Syracuse University Press.

Zhang Baijia. 2001. "The Changing International Scene of Chinese Policy Towards the United States, 1954–1970." In Robert S. Ross and Changbin Jiang, eds., *Re-Examining the Cold War: U.S.-China Diplomacy, 1954–1973*. Cambridge, Mass.: Harvard University Asia Center.

Zhao, Quansheng. 1996. *Interpreting Chinese Foreign Policy: The Micro-Macro Linkage Approach*. Hong Kong: Oxford University Press.

Zhao, Quansheng, and Guoli Liu. 2010. "Beijing's Shifting Positions in the New Era of Cross-Taiwan Strait Relations." In Wei-Chin Lee, ed., *Taiwan's Politics in the 21st Century: Changes and Challenges*. Singapore: World Scientific Publishing.

Zisser, Eyal. 2002. "Israeli-Syrian Peace Negotiations (December 1999–March 2000): A Missed Opportunity?" In Joseph Ginant, Edward J. Perkins, and Edwin G. Corr, eds., *The Middle East Process: Vision versus Reality*. Norman: University of Oklahoma Press.

INDEX

www.ingramcontent.com/pod-product-compliance
Lightning Source LLC
LaVergne TN
LVHW091517200926
848483LV00005B/435/J

* 9 7 8 0 8 1 2 2 4 4 9 8 4 *